ACTION
OF ICE ON ENGINEERING
STRUCTURES

by

K. N. KORZHAVIN

Books for Business
New York-Hong Kong

Action of Ice on Engineering Structures

by
K. N. Korzhavin

ISBN: 0-89499-171-X

Reprinted from the 1971 edition

Books for Business
New York - Hong Kong
http://www.BusinessBooksInternational.com

ACTION OF ICE ON ENGINEERING STRUCTURES

Vozdeystviye l'da na inzhernernyye
sooruzheniya, Izdatel'stvo Sibirskogo
Otdeleniya Akademii Nauk SSSR, (Publishing House of Siberian Branch of
USSR Academy of Sciences), 1962, 202 p

K.N. Korzhavın

CONTENTS

SECOND PART

Survey of Physico-Mechanical Properties of River Ice

THIRD PART

Pressure of Ice on Structural Abutments (Supports)

FOURTH PART

Pressure of Ice on Structures of Considerable Extent,
and Other Cases

FIFTH PART

Natural Methods of Determining Ice Pressure on
Structural Supports

SIXTH PART

Experience Gained in Determining Actual Pressure of
Ice on Structures by Use of Kinematic Method

SEVENTH PART

Passage of Ice and Size of Ice-Admitting Openings

EIGHTH PART

Findings and Conclusions

PREFACE

The tremendous potentialities of the building of communist society launched in the CPSU program and in the resolutions of the 22nd CPSU Congress require an even broader development of industry and transport in the country.

The richest hydropower resources of the Angara and Yenisey Rivers will be exploited; we have created a unified deep water system of waterways in the European sector of the Soviet Union, we have greatly expanded the network of railways and highways and we have implemented a vast program on the irrigation of lands and water supply to cities and towns. In connection with this, we will erect many large engineering structures (dams, hydraulic power stations, bridges, docks, water intake structures etc.), which are exposed to some extent or other to the action of ice in the winter and particularly in the spring period. Consideration of the ice action assumes special significance for the regions of Siberia and the Soviet Arctic; for the rivers there, we typically have a significant thickness and strength of ice, plus an intensive ice breakup.

At the same time, regarding a number of questions on ice action, there are still no generally recognized solutions. The physico-mechanical properties of ice up to the period of rivers' breakup have not been studied as they should have been; various opinions exist on the effect of the structures in plane and profile, the role of friction in the process of ice action on the structures. There have been no methods for determining the dimensions of the ice-passing openings and in effect no convenient methods for determining the actual pressure of ice under natural conditions.

In connection with this, the need has matured for refining the physical pattern of the phenomenon, to analyze the accumulated materials and to advance certain new, more substantiated methods for determining the ice loads on the engineering structures.

The report has been written on the basis of generalizing the literature data and also on many years' research conducted by the author on the questions of ice action on structures under the conditions of the ice breakup in Siberian rivers, conducted from 1934-1961 (initially in the Novosibirsk Institute of Engineers of Railway Transport, and from 1952 in the Ice Thermics Laboratory of the Transport-Power Engineering Institute of the Siberian Branch of the USSR Academy of Sciences).

The author expresses deep appreciation to Associate Member of USSR Academy of Sciences, V.V. Zvonkov, Prof. A.Ya. Aleksandrov and Prof. K.K. Yakobson for their unflagging participation in the work and also to the staff members at the Institute, especially I.P. Butyagin, Y.Ya. Kazub and V.K. Morgunov.

INTRODUCTION

The construction organizations of the USSR are confronted with the tremendous tasks in planning and erecting many of the largest engineering structures and are exhorted to meet the ever-growing requirements of the national economy of the country.

Many of these structures--the large bridges on the new railroad and highway routes, dams at unique hydroelectric stations, the water intake structures--are subjected to some degree or other to the action of the ice sheet both in the winter and par-ticularly in the spring period.

The allowance for the effect of the ice cover on the struc-tures is necessary, since many USSR rivers are typified by an appreciable ice thickness and intensive spring ice debacle. Par-ticularly severe are the ice conditions in the rivers of the Soviet Arctic and Siberia, having a permanent ice pack of con-siderable thickness and an intensive spring breakup, transpiring at uniquely high levels with the formation of ice jams.

It is possible to differentiate the following forms of ice action on engineering structures:

a) static pressure, developing during the formation of an ice sheet or during its expansion as a result of abrupt variations in air temperatures in the winter season;

b) dynamic pressure of ice pack in period of spring and fall ice motion at encountering individual abutments or struc-tures of large size (dams, arches, embankments, ice-protective walls, etc.); and

c) abrasive effect of ice pack during ice debacle.

In addition to the horizontal forces the ice pack can cause vertical forces at meeting with inclined edges, at fluc-tuation in water level, or at falling of floes into the down-stream water below the structure.

Whereas in winter, the structures undergo relatively slight pressure of the ice sheet (while in fall, they experience a slight dynamic pressure), during the spring ice debacle, one can observe the dynamic action of large ice fields at times drifting with high speed. The forces of interaction originating at this time can be quite significant and can lead to serious damages to (or even the destruction of) the structure, as was sometimes observed in the practice.

Let us point out that the static pressure of ice has importance chiefly for structures of considerable extent (dams, dock structures, ice-protective walls), whereas the dynamic action is most hazardous for the individually standing structures (bridge abutments, piers of overflow weirs, beacons, water intake structures etc.).

For providing strength and stability to the structure, of great importance is the division of the bridge openings into spans or dimensions of the ice-discharging openings. With narrow spans, it is possible to have the stopping of the ice fields and even the formation of ice jams at a structure, with very serious consequences.

In the planning, construction and operation of a structure on rivers with severe ice conditions, it is mandatory to provide:

a) strength and stability of all structural parts, with consideration of the static and dynamic action of ice;

b) jam-free passage of the ice fields in the bridge opening or other structure; and

c) protection of structure from damages at point of floe's impact (from punching out of masonry, facing rocks, from forming cracks etc.).

From time immemorial, the solution of the problems formulated has attracted the attention of many researchers and engineers, (chiefly Russian and Soviet), having undertaken full-scale observations, experimental and theoretical studies of the question. The paramount role of the Russian and Soviet school in the solution of the problem of combatting the ice is established by the fact that the native scholars investigated the problem on the basis of the deepest possible penetration into the physical nature of the phenomenon.

Generally speaking, there are few foreign studies conducted in the field of combatting ice and they are marked often by a schematization of the phenomena, without reflecting the actual state of affairs. In connection with this, it is not surprising that a bridge across the Connecticut River at Brattleboro (U.S.) has been swept away by ice 8 times in the last 120 years, while

Columbia Dam in Pennsylvania has been broken down 4 times in the last 18 years from the impact of ice floes, where the length of the damaged section attained 270-122 m [1].

In the USSR, such cases of repeated destruction of major structures have not occurred in spite of the more severe ice conditions in the rivers of our country.

Let us examine the degree of development of the problems formulated above.

1. The initial systematic task force studies of the winter conditions were conducted on the Yenisey River under the supervision of Prof. Ye. V. Bliznyak [2]. Extensive investigations were then made on the Volkhov and Svir' Rivers under the guidance of V.M. Rodevich, then on the Angara River under the leadership of V.M. Malyshev and on the Lower Volga under the guidance of B.V. Polyakov [3]. In addition, those occupied with studying the winter regime of the USSR rivers included V.Ye. Timonov [4], V.T. Bovin [5], F.I. Bydin [6,7], Ye. I. Ioganson, M.F. Tsionglinskiy, D.S. Artamov [8], N.D. Antonov [9] and many other researchers, having achieved significant results.

Much was accomplished, particularly in the Soviet period, when an appreciable role was played by the State Hydrologic Institute (SHI) (organized in 1919), the All-Union Arctic Institute, Main Directorate of USSR Hydrometeorological Service (founded in 1929), and also the large planning and research organizations (GIDEP, Gidroproyekt, Giprorechtrans, Soyuztransproyekt etc.), having accumulated the most valuable data on the winter conditions in many USSR rivers.

However, the study of the USSR river conditions can by no means be considered completed, especially in respect to the small and medium-size rivers, or to certain large rivers in the Soviet Arctic and Siberia in their lower course. For instance, we have extremely insufficient data on the sizes of floes and the speed of their movement at various periods of the ice debacle, and also concerning the fluctuations in the actual depth of ice pack by width and length of river up to the time of ice breakup. There are few data on the conditions of the formation and breakup of ice jams, concerning the thermal regime of the ice pack, on the extent of the ice pack's weakening up to the period of breakup under the effect of the spring thawing processes. All of these questions, extremely significant for the planning of measures on counteracting the ice difficulties, are still waiting to be solved

2. The mechanical and elastic properties of ice are related complexly with many factors, above all with ice temperature, rate of loading, dimensions of sample, structure, and orientation of pressure relative to the crystallization axis.

Therefore, although many scientists and specialists have been engaged with studies of the ice's mechanical properties (including B.P. Vasenko [10], S.O. Marakov [11], A.N. Krylov, B.P. Veynberg [12,13], V.N. Pinegin [14], B.N. Sergeyev [15], M.M. Basin [16], V.S. Nazarov [17], F.F. Vitman [18], K.N. Korzhavin [19,20,21,22], B.G. Korenev, S.M. Izyumov [23], B.D. Kartashkin [24], A.R. Shul'man [25], V.P. Berdennikov [26], N.N. Zubov [27], Yu. N. Neronov, [28], N.A. Tsytovich [29], I.P. Butyagin [30, 31,32], K.F. Voytkovskiy [33], S.S. Vyalov [34,35], V.V. Lavrov [36], B.A. Savel'yev [37] and others), nevertheless, the question has not yet been adequately studied.

There are little data on the actual strength of ice during the spring ice debacle, and even less information on the strength of ice under natural conditions of a drifting ice pack. The large-scale tests dictated by the heterogeneity of the ice were undertaken in the USSR only by certain researchers, specifically V.N. Pinegin, M.M. Basin, I.S. Peschanskiy [38], the author and by I.P. Butyagin.

At introduction of an abutment into a floe, we find the phenomenon of local crushing, significantly affecting the forces of interaction between the abutment and the floe. However, the effect of local crushing and also of the loading rate for ice strength has not yet been clarified in the merited manner. The ultimate shearing and tensile strength of natural ice has also been insufficiently studied. For the Poisson coefficient of river ice, there are only isolated determinations; very different values are recommended for the modulus of elasticity.

All these factors have required the undertaking of further systematic studies.

3. The value of dynamic ice pressure on the structures is also a complex function of a number of factors, among which we can name the dimensions, form and speed of floe movement; the physico-mechanical properties of ice; dimensions, form and material of abutment body, presence of wind, saturation of river surface with ice, and the form of the flow's water surface.

Even a short list of the factors is indicative of the great complexity of a theoretical review of the problem formulated and the desirability of undertaking experimental and natural observations. However, the difficulty of organizing the natural observations has led to the development of a series of theoretical methods based on some given systematization of the phenomenon. This study has attracted the attention of eminent scientists, particularly Academician G.P. Perederiy [39], Prof. L.F. Nikolai [40] and N.M. Shchapov [41].

The development of methods for determining the ice pressure has also been the topic of reports by N.A. Rynin [42], A.N. Komarovskiy [1, 43], F.I. Bydin [44], A.A. Dubakh [49], B.V. Zylev [45], I.S. Peschanskiy [38], A.I. Gamayunov [46], P.A. Kuznetsov [47,51], N.N. Petrunichev [48], Ye. V. Platonov [50], I.M. Konovalov [52] and the author [53,54,55,56,57,58].

It should be pointed out that we studied in more or less detail only the question of the forces of interaction developing at encounter of an abutment with a small floe, the entire kinetic energy of which is expended in disruption. The other, actually more interesting problems of the encounter of large floes with a structure (when the floe is either cut through by the abutment or sheared) have not yet been solved satisfactorily. The procedures suggested by the authors overlook the phenomenon of local crushing, speed of floe's movement, form of abutment in plane and in connection with this, they can be adopted only as a first approximation.

The amount of the static pressure of ice on structures has also been studied by a number of authors (A.I. Komarovskiy [1], V.P. Skryl'nikov [59], V.T. Bovin [5], N.I. Anisimov [60], I.G. Aleksandrov, K.M. Dubyaga, B.Yu, Kalinovich, S.S. Golushkevich [61], N.N. Petrunichev [62] et al.), most of whom recommend using the familiar proposals made by N. Royen [63]. However, as was shown by B.V. Proskuryakov [64], in the derivation of the Royen formula, significant inaccuracies were permitted as well as poorly based assumptions, which permitted the introduction of important suggestions in the given field.

In connection with this, the problem of the further development of methods for determining the pressure of ice on structures (both theoretical and full-scale) continues to remain urgent.

4. The question of the minimally permissible size of structural span, proceeding from the conditions of ice passage, has been studied little, while the requirements of the practice are solved on the basis of work experience with the existing structures.

5, An efficient form of an abutment in plane and profile and its effect on the forces of interaction have not yet been clarified to the extent deserved. As a result of this, the Technical Specifications on Bridge Planning (TSBP) recommended the adoption of the same value of ice pressure for all abutments with a cutting edge, apart from any dependence on the abutment's form in plane. The effect of the slope of the ice-cutting edge was not evaluated quite properly either.

We should thus consider the extent of the question's study as still inadequate; at times, this leads to an obviously inefficient expenditure of material resources. Thus, in the planning of abutments to bridges across rivers with severe ice breakup conditions, use is sometimes made of abutments with a sloped ice-cutting edge, wherein the amount of slope is chosen more or less arbitrarily, without enough substantiation. In this way, there is a significant increase in the scale and cost of operations even when there is no necessity for this.

On the other hand, we are aware of the fact of serious damage to several supports of a large bridge (having been built in the south of the USSR across a broad waterway) by an ice field, transported by the wind and current. In this case, the amount of ice pressure was obviously underestimated by the planners and builders.

In connection with this, we should consider a further study of the question as quite necessary.

Content of Study

As a result of the research conducted, in the present report:

a) the author clarifies certain features of the spring ice debacle in the USSR rivers with utilization of natural observations of the breakup of many rivers in Siberia;

b) a clarification is given of the actual operating pattern of the structural supports' ice-cutting edges at various periods of the ice passage, and cases of the disruption of floes are classified. This section utilizes our natural observations conducted from 1933-1961 in the Siberian rivers.

c) recommendations are made of the calculation values for the strength of ice (in the period of river breakup) in respect to compression, crushing, bending, shearing and rupture.

In addition to the literature data, and above all I.P. Butyagin's reports [30,31,32], use has been made of extensive experimental research conducted by the author in establishing the strength of ice directly in the period of spring ice debacle and also in clarifying the effect of the rate of loading, ice temperature, local crushing and orientation of pressure relative to the crystals' axis.

An explanation is given of the very significant role of the specimen's dimensions, phenomenon of local crushing, and also of the rate of loading; a substantiation is given for the computed values of the ultimate strength under actual conditions of the abutments' functioning;

d) an explanation is given of the effect of an abutment's form in plane on the resistance of the ice pack based on experiments undertaken based on punching (out of the river ice) core samples of varying form, and also based on theoretical developments.

It is emphasized that the abutments with a sharper outline of cutting edge require much lower forces for penetration into the ice.

We recommend introducing into the calculations the factor of the abutment's form and we have suggested its numerical values;

e) we have suggested new methods for determining the ice pressure on the structural supports during the cutting, stopping and splitting of a floe, with consideration of the effect of the region's climatic features, the abutment's form in plane and profile, phenomenon of local crushing and rate of floes' motion.

The calculated relationships are supported by theoretical concepts and lead to a simple, compact form, convenient for practical application;

f) a clarification is given of the role of the ice cutting edge's inclination as a factor facilitating the disruption of the ice sheet.

The author indicates the possibility of the breakup by the inclined edge of an ice apron (starling) not only by means of bending but also by shearing. It is shown that destruction by shearing is widespread and a method is recommended for determining the ice pressure both in this case and in the case of bending. Attention is diverted to the fact that the inclined ice-cutting edge should be used only in extreme cases after careful justification. Concepts are presented in favor of the idea that in place of expensive abutments with an inclined edge, one can sometimes apply vertical supports with a sharper outline of the cutting edge;

g) a simple and sufficiently accurate method is suggested for determining the actual ice pressure on the structural supports in natural conditions based on an analysis of the conditions of motion of a large floe, being cut through by supports. It is demonstrated that at disruption of a large ice field by structural supports, a significant role is played by the additional force of the flow formed from friction on the lower rough surface of the floe. The author takes up the question of the rate of a floe's motion under condition of complete ice breakup and dispersed ice.

Experience gained in using this method is discussed, and a comparison is made of the actual pressure of ice with the calculated values, based on various methods; and

h) a method is recommended for determining the minimal permissible bridge span or opening in a hydraulic engineering structure, proceeding from the condition of the unjammed passage of ice.

The present report does not profess to exhaust all the facets of the problem; in this connection, we have indicated in the conclusion some areas where further research into the problem is needed ([Note]: Static pressure of an ice pack has not been reviewed in great detail in the present report, since it has significant value mainly for structures of great extent (dams), usually having considerable dimensions and in connection with this, withstanding successfully the ice pressure at observance of the proper conditions).

FIRST PART

NATURE OF INTERACTION OF ICE AND STRUCTURES

Chapter I

CERTAIN FEATURES OF SPRING ICE DEBACLE (BREAKUP) IN USSR RIVERS

1. Basic Concepts

The forces of interaction originating at encounter of an ice field with a structure are particularly significant in spring, in connection with which the features of ice breakup in the USSR rivers deserve a more detailed clarification.

In the vast territory of the USSR, we find zones which are quite diversified in their physico-geographic and climatic features, and which also determine the difference in the winter regime of the corresponding river sectors. The conditions of spring ice debacle are closely related with the local features of the region under consideration and specifically with the thickness and strength of the ice pack up to the time of breakup; the direction of river current relative to the points of the compass; the amount of precipitation falling as snow; and the variation in air temperature in the spring period.

The opening of rivers flowing from north to south begins from the lower reaches and the ice, gradually breaking away from the edge of the ice pack, is floated downstream. In this case, the ice strength, as a result of a rise in the air temperature, effect of solar radiation and the entrance of melt water into the river, decreases considerably, and in these rivers the spring ice breakup usually proceeds relatively smoothly.

Other conditions of ice breakup are created in the rivers flowing in a meridional direction from south to north (many rivers in Siberia and the Soviet Arctic). The melting snow, commencing in the upper river reaches situated farther south, causes a rise in water and ice breakup. The flood water wave, shifting down the river, encountering on its way the ice which is still fairly permanent, nevertheless breaks it up; for this reason, in these instances one should reckon with the possibility of considerable ice strength during the spring ice breakup period. In addition, the drifting ice masses often encounter an ice pack not yet broken up, creating favorable conditions for the formation of ice jams and catastrophic rises in water level, with which the ice breakups of many Siberian rivers are accompanied. These are the most serious conditions of ice debacle, proceeding violently, at high water levels, with numerous manifestations of the ice's destructive activity.

Finally, a serious effect on the nature of the river's opening is exerted by the amount of snow in a basin and the pattern of air temperatures during the spring period. For instance, in the Transbaikal regions (and partly in the Far East) owing to the inadequate precipitation in the form of snow, the ice breakup proceeds at low levels, more or less smoothly, in spite of the considerable depth of the ice sheet. The rapid warming, encompassing a considerable part of the river basin leads, on the other hand, to an abrupt rise in the water level, an early breakup of ice which is still fairly strong, and it intensifies the effect of the ice sheet on the structures. A lingering spring promotes the gradual thawing of the ice pack even prior to the breakup and the smooth ice debacle.

Summing up, we can conclude that in the USSR territory, we have the possibility of observing: a severe ice debacle--in the rivers of the Soviet Arctic, in the middle and lower reaches of the large Siberian rivers; an intensive breakup in the upper reaches of the large Siberian rivers, in the rivers of the Urals and in the north of the USSR European sector, with a weak ice debacle in the rivers of the southern and western sectors of the USSR.

For a more detailed description, let us examine certain features of the spring ice breakup: a) of large rivers, flowing from south to north (in the example of the Siberian rivers); and b) of large rivers flowing from north to south (in the example of rivers in the European part of the USSR).

As a result of their reduced practical significance, the features of ice breakup in small rivers are touched on only briefly below.

2. Features of Spring Ice Debacle in Large Siberian Rivers

The territory of Siberia also includes a number of zones quite diversified in their physico-geographic features; this condition also establishes the conditions of the thawing and freezing of the Siberian rivers.

The southeastern part of Siberia is occupied by the Altay mountainous country and by the Central Siberian Upland, watered by mountain rivers with a predominant role of glacial and snow feeding. For the southwest of the territory under consideration, occupied by vast steppe areas situated in the zone of deficit water supply, we typically have a few rivers with scarce water flowing slowly through the West-Siberian Lowland. The north is occupied by forest-steppe and taiga-swamp zones, then transforming to tundra.

In the territory of Siberia, we find both the rivers which do not have intensive spring ice breakup (Transbaikal, Minusinskaya Basin, certain rivers in Northern Kazakhstan), and also the rivers with quite severe spring ice breakup, progressing tempestuously at high water levels with the formation of water buildups under snow, and ice jams.

Still more serious are the ice debacle conditions in the lower stretches of the great Siberian rivers, bearing a huge mass of water into their lower reaches still not freed of ice.

The basic features of the spring ice breakup reduce to the following:

1. Thickness of ice sheet in the Siberian rivers up to the time of opening will fluctuate from 0.9 to 1.7 m, reaching a maximum of 1.75 - 2.8 m. We should note particularly that sometimes there does not occur an appreciable reduction in ice thickness up to the period of ice breakup (as a result of the thawing process).

-17-

2. By the period of the spring passage of ice downstream, the ice sheet becomes very heterogeneous and is cut by a series of fissures. Its lower surface is covered with hollows of an undulant nature, having a depth up to 10-15 cm. Let us point out that the thickness of the ice pack up to the breakup time will fluctuate within considerable limits in the same river sector.

3. The breakup of the Siberian rivers is determined chiefly by the mechanical effect of the high water wave, in connection with which the strength of the ice pack will still remain considerable during the ice passage period.

4. Ice in many rivers of Siberia will move from south to north and ice jams are easily formed, meeting along their route the ice sheet which is not yet broken up. In connection with this, the highest levels in the year either coincide with the ice debacle or are close to it.

5. The first movement sometimes occurs at considerably higher speeds than is usually assumed (0.8 - 1.2 m/sec).

6. The area of individual floes attains very considerable extents, wherein the motion of large floes is also possible in the period of scattered ice.

7. The maximum speed of floes' movement varies greatly depending on level of ice breakup, quantity of drifting ice and other local conditions; for floes of small size (10 - 1500 m^2), the speed is usually 1.5 - 2.5 m/sec.

The large floes ($\Omega > 100,000$ m^2) in the period of total ice passage will move with a speed of 0.5 - 1.5 m/sec, while the ice fields will move with a speed of 0.6 - 0.8 m/sec.

3. Features of Spring Ice Debacle in Certain Large Rivers in European Sector of the USSR, Flowing from North to South

In the given instance, we will conduct the explanation of the spring ice breakup in the example of one river having an extensive watershed, and situated in regions with diversified climatic conditions.

From the viewpoint of spring ice breakup, the basin of a river can be divided into two zones: an upper zone, characterized by a relatively severe winter, a considerable thickness of ice, broken up in spring by the passing high flood water, and a lower zone, with a shorter and milder winter, with thin and relatively weak ice.

The features of the spring ice breakup in the river under consideration reduce to the following:

1. The shorter and warmer winter leads to the situation that the ice pack has a lesser thickness, which up to the spring breakup period decreases perceptibly (by 50 - 30%) owing to the spring thawing processes.

2. The strength of the spring ice in the southern sectors of the river up to the ice breakup period also decreases markedly owing to the weakening of the bond between individual crystals.

3. Ice strength in the river headwaters located farther north is high, which causes the formation of floes of larger dimensions than in the lower reaches.

4. The opening in the lower course of the rivers occurs gradually by way of the breaking away of the ice (in a river sector with a length of 50 - 100 km), which then drifts smoothly downstream.

5. The opening of the upper reaches of the rivers is accompanied by an ice breakup of great force and sometimes by ice jams, appearing more often during a prolonged spring with alternating periods of cooling and in the tapering zones of backwater from the hydraulic structures.

6. Natural observations of the spring breakup of large rivers in the USSR European sector flowing from north to south indicate that with respect to the action of ice, the abutments of the structures are under much better conditions than on the rivers of the Soviet Arctic and Siberia.

4. Variation in Ice Breakup Conditions Caused by Control of Rivers

The ice conditions in a supported channel differ considerably from the usual ones in connection with the abrupt increase in depths, reduction in current velocities, change in temperature regime of the reservoir and the increase in the wind's role. The thickness of the ice sheet varies relatively little, increasing somewhat in connection with the decrease in current velocity and decreasing considerably at the dam itself, with the upwelling of the lower, warmer water layers.

The clearing of the large reservoirs of ice occurs later (on an average by 10 - 12 days) as compared with a river in a natural state, although in certain large reservoirs of the USSR, there is a lag in the periods of breakup in individual years of only 2 days, while in other years, the lag can extend to 22 days.

Sometimes ice passage downstream does not occur at all and the ice thaws <u>in situ</u>, as has been noted repeatedly in the upstream waters (<u>backwater</u>) of certain reservoirs in the Soviet Union. It is interesting to note that in one of the ice debacles, the ice field in the backwater of a large reservoir remained in place, in spite of the fact that all the sluices of the spillway part of the dam were opened, and only after 4 days did the field pass over the dam. N.A. Girillovich [65] suggests that ice passage over a dam is possible only at current velocity greater than 0.4 - 0.5 m/sec, while at lower values, ice passage is possible only under the effect of wind.

The strength of the ice sheet on a reservoir decreases considerably as a result of a delay in the opening of the river and the effect of a higher water temperature. D.N. Bibikov and N.N. Petrunichev [66] recommend a determination of the decline in ultimate compression and shearing strength of ice in the pre-ice passage period based on the relationships:

$$R'_{compr} = R_{compr} - az,$$

$$R'_{sh} = R_{sh} - bz$$

$$\text{at } a = 1.5 \text{ kg/cm}^2 \text{ per day,} \quad R_{compr} = 35 \text{ kg/cm}^2,$$

$$b = 0.3 \text{ kg/cm}^2 \text{ per day,} \quad R_{(sh)earing)} = 10 \text{ kg/cm}^2,$$

$$\text{and } z = \text{number of days.}$$

We can thus consider that in an opening delay of 10 days, the limits of the compressive and shearing strength of ice will decrease significantly and will comprise 0.70 - 0.60 of the ice strength up to the beginning of the spring ice debacle in the uncontrolled sectors of the river. In designing the piers of the overflow (spillway) dams, one should proceed from large dimensions of the floes, slower current speed and a reduced ice strength (by 25 - 30%).

5. Features of Spring Ice Breakup on Small Rivers

On the small rivers, the spring ice passage has unique features, among which we can indicate the following:

a) in most cases, the small dimensions of the basin lead to an almost simultaneous opening of the river over the entire length, in connection with which there occurs a rapid although

ephemeral rise in water and coincidence of the highest levels
with the ice breakup;

   b) the small rivers usually break up sooner than the
large ones; it is therefore necessary to reckon with a somewhat
greater strength of the spring ice; and

   c) the smaller width of the rivers causes the appearance
of floes having small dimensions.

   One therefore should consider that the spring ice passages
on the small rivers do not represent a great danger for the major
structures.

   However, the temporary structures often built along·the
small rivers (wooden bridges, crib and earthen dams) are often
damaged and at times even destroyed by ice, chiefly from the
formation of ice jams, a rise in water level, and underscouring
of the structure. One also finds damages to wooden supports
by impacts from floes during the ice passage period. Therefore
during the planning of structures on small rivers, the water-
conduit openings and the design of supports should be adopted
with consideration of the features in the river's ice regime.

   Thus in many rivers of the USSR, the spring ice breakup
comprises a serious factor which must be taken into account in
planning the structures.

   In all cases, therefore, it is quite necessary to make a
careful study of the winter regime of rivers in the region of
planned structures for obtaining information (the actual thickness
and strength of ice, dimensions of floes and speed of their movement, conditions of ice jams' formation, fluctuations in water
level), necessary for an actual consideration of the action of
the ice pack. Failure to comply with this requirement can lead
to damage or even destruction of structures by the action of ice,
as has repeatedly taken place in practice.

   Certain cases of damage and destruction of structures by
the ice pack have been presented in the next chapters.

Chapter II
NATURE OF ICE INTERACTION WITH STRUCTURES
1. Basic Concepts

The efficient planning of the supports of bridges and of hydraulic engineering structures on rivers with severe conditions of ice passage can be successful only with a combination of the methods of theoretical analysis with a broad generalization of the experience gained in the operation of the pertinent structures. Such observations are still few in number.

From 1931-1961, the Novosibirsk Institute of Railway Transport Engineers (NITRE), and from 1951 the Laboratory of Ice Thermics at the Transport-Power Engineering Institute of the Siberian Branch of the USSR Academy of Sciences (TPEI), under our supervision, conducted observations of the effect of ice on structures under the conditions of ice passage in the rivers of Siberia.

In 1940, A.A. Trufanov and G.N. Petrov [67] undertook laboratory studies on the question of an efficient form of starlings (ice aprons) for wooden bridges, while in 1946, A.I. Gamayunov [46] conducted observations of the functioning of bridge supports during the spring ice breakup on the Dnieper River at Kiev. In 1954, G. Karlsen and N. Streletskiy [68] published some interesting data on the operation of the starlings of the hipped-cover type. In 1948, A.S. Ol'mezov and G.S. Shpiro [69] described the function of the supports to one large bridge having been severely damaged in its abutments during the spring ice passage.

Many worthwhile ideas on the given question are contained in the reports by G.P. Perederiy [39], L.F. Nikolai [40], A.N. Komarovskiy [1], N.A. Rynin [42], Ye.Ye. Gibshman [70], Ye. V. Boldakov [71] and certain other authors.

Let us review some possible systems for the breakup of an ice cover.

2. Work of Bridge Abutments During Movements of Ice

During the passages, an ice field breaks away from the ice sheet and is brought into motion under the effect of frictional forces of water flow on the lower rough ice surface, or of the air flow on the upper, often hummocky, surface of the floe, or even from the combined action of the current and the wind.

The ice field which has broken away begins to increase the rate of its motion up to a certain critical value, corresponding to the given actual conditions (area of floe, its roughness, thickness, slope of flow surface, velocity of current).

The abutments encountered by the moving ice pack naturally delay the motion of ice. The velocity of the flow decreases while in certain cases, we can note the stoppage of the ice field at the abutments; we observed this condition several times.

Thus, for example, in April 1935 during the movements of ice on one of the Siberian rivers, an ice field with a width of around 160 m became jammed in one of the bridge spans (more than 100 m in width) and stopped. Even though a cooling occurred (temperature fell to -10°C), a new movement of the ice took place during the night, having ended in the stopping of another field (900 x 400 m) at the third pier of the bridge.

In April 1950, an ice field measuring around 600 x 300 m was halted by four bridge piers and stayed in such a position for around 2 days. Often the countereffect of the piers proved insufficient and the field continued to move, dividing into strips drifting alongside. In this case, the cruising rate of the flow naturally decreased.

In a case described by A.S. Ol'mezov and G.S. Shpiro [69], an ice field, having moved with a velocity of 0.3-0.4 m/sec, upon encounter with the pier also became cut into strips with a velocity of 0.01-0.02 m/sec and, moving for 50 m, came to a stop. A.I. Gamayunov [46] observed during the 1946 ice passage how a field with a width of 100-150 m and a length greater than 1 km became cut through continuously by a pier into two parts until a remnant of the floe with a length of 150-200 m broke up. The velocity of such a floe was about 0.5 m/sec.

In this manner, a large ice field, having a significant kinetic energy reserve, can be halted by the piers only after the entire force has been expended in the work of breaking up the ice and in the work of frictional forces on the frontal and lateral surfaces of the pier.

The crushing process transpires as follows: an ice field, encountering the pier face, experiences the local crushing phenomenon, accompanied by appreciable plastic flows. In proportion to the penetration of the pier into the ice, there is also an increase in the force of reciprocal action between the support and the floe, which can cause crushing, or breaking of the floe, or even its destruction from cutting under the effect of the vertical component of ice pressure (Fig. 1). Breakdown of the ice sheet by fracture occurs at a very short distance from the pier, not exceeding 3-6 thicknesses of the ice.

The diagram brought to the reader's attention indicates the validity of such a conclusion. Figure 2 illustrates the moment of breakdown of a low-strength ice field during the 1935 ice-out in Siberia. Breakdown took place owing to the shearing action. The ice was forced upward and crumbled into a heap of white snowy typically crackling needles. At this time, the total contact of

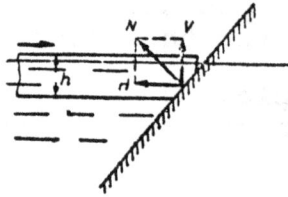

Fig. 1.   Action of Ice on Sloped Face of Structure.

the ice sheet with the pier along the entire perimeter of the
ice apron occurred. In the photograph (Fig. 3), we have shown
the more permanent ice, not having broken up into separate needles
but into chunks of small dimensions, owing to shearing or bending.

A similar pattern was observed in the breakup of an ice pack
by icebreakers. The nature of the disintegration during the
climbing of the icebreaker onto an ice field is portrayed in Fig.
4. At this time, initially cracks appeared perpendicular to the
vessel's side, having divided the strips of ice, which then showed
transverse cracks [73]. According to the observations of the
Scientific-Research Institute of Navigation and Shipbuilding conducted on ice with a thickness of 60 cm, the transverse cracks
were observed 6-7 m from the ship's side.

Fig. 2. Disintegration of Ice Sheet by Means of
Shearing Action.

The appearance of radial cracks, along which the mutual displacements of adjoining sectors of the floe occur in vertical
planes, is explained by the complex form of the submerged section
of the icebreaker. Certain sectors of the ice sheet meet with
parts of the hull which project most extensively, causing the lower
ing of these sectors and their separation by shearing away from the
main ice mass. After this, the adjoining sectors come into con-

tact with the hull; these sectors once again are separated from the contiguous ones by means of shearing, etc.

Fig. 3.   Destruction of Ice Sheet Owing to Bending
Action

Without doubt, a similar pattern is found when the structural support meets a drifting ice floe. The more or less complex shape of the ice apron's cutting face leads to the situation that part of the ice field is raised by the sloped face of the ice apron and is then torn from the remaining part of the ice pack owing to the shearing or bending action.

In this way, the processes of ice field crushing during shifting develops as follows: meeting the sloped ice-cutting pier face, the ice field attempts to climb onto it, which evidently also takes place with an ice apron of great width, approaching the width of the ice field. However, since the pier's width is usually slight, only a minor part of the floe shows a tendency to climb up, and moreover this part is connected directly with the surrounding ice pack. The great actual weight of the ice field and the relatively low shearing and breaking strength of the ice (in the spring period) favor the situation that the climbing part separates from the remaining ice field both along its length and width. In general, it should be noted that the phenomenon of the floes creeping onto a sloped starling is closely related with the strength of the floe itself, wherein dependence exists: the more suitable the floe, the higher it will climb, and the larger the dimensions of the part of the field which is being raised.

In the spring of 1935, in Siberia on the identical day and at the same hour, with weak ice, practically no climbing took place, while at encounter of a pier with a strong floe, over which a winter road had traversed the river, the climbing was considerable.

The part which climbed up broke into small bits (of the order of 0.2 - 1.0 m^2), which slid down under their own weight onto the ice sheet which had again advanced on the pier, or they (the bits) sank under the sheet by sliding along the pier. The presence of individual chunks of ice between the floe and the pier and also the actual process of breaking up the ice field usually prevents the ice sheet from resting tightly against the pier along the whole perimeter of the sloped ice-cutting face.

In connection with this, the simultaneous pressure of the ice sheet (at some appreciable strength of it) as a rule could not occur over the entire perimeter of the sloped star-ling. During the movement of large ice fields over the river's entire width, and also in case of a steep or straight star-ling, the closeness of contact of the ice field to the abut-ment doubtless increases; nevertheless, the pressure distri-bution over the entire perimeter of the abutment no matter what the thickness of the ice, is impossible.

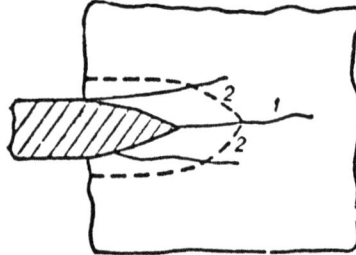

Fig. 4. Nature of Disruption of Ice During Encounter
with Support: 1 - leading crack; and 2 - lines
of disruption.

At a first approximation, the looseness of the ice field's
contact with the support can be taken into account by multiply-
ing the support's width b_o times the factor of contact completeness k, as was suggested in our report [54]. In this manner, the width of the support absorbing the simultaneous pressure of the ice can be assumed to equal:

$$b = b_o k.$$

According to our observations, for floes of average
strength, the approximate k-value can be assumed to fall within
the limits of 0.4 - 0.70.

3. Function of Bridge Supports During Period of
Total Ice Breakup

In the period of complete ice passage, the starling acts
as a wedge which divides the mass of moving ice, to the extent
that local conditions permit this (dimensions of floes, concentration of ice on the river surface, form of starling in plane
and profile, current velocity, etc). This process develops in
different ways at meeting of the starlings with the large and
small floes.

The small floes, striking the starling (ice apron) and
not having the opportunity to move aside, climb onto it, sometimes tilting, again plunging in the water, and often turning
over several times in the water. At this time, the support sus-
tains impacts from individual floes at various points and at
different times.

As we have observed repeatedly, the large floes often split apart. Upon meeting a large floe, the support began to penetrate into the ice, at which time the force of interaction between the support and the floe gradually increased. At a certain moment, the force attained a magnitude sufficient to split the floe completely, or partly, with the formation of a crack over some of the floe's length.

Thus, we observed fissures with a length up to 5 - 10 m and more, where their direction most often deviated from the support's axis by 10 - 30° to one side or the other. At times, the crack had a zigzag shape in plane, indicating the possibility of the opening of old fissures having cut into the ice field even prior to encountering the bridge support. At the support itself, we noted the disruption of the ice field at the rising of the floe's edge onto the sloped ice apron. After the shearing and breaking off of the floe's edge, pressure on the support dropped, as could be judged by the closing crack (lead). A new opening and lengthening of the crack then took place.

In this manner, the pressure of ice on the support at disruption of one floe occurs irregularly, by jolts, as is also confirmed by the vibration of the support which took place.

The increase in the force of interaction between the support and the floe occurs until the support has cut into the floe for its entire length. For the further advancement of the support into the ice, a constant, almost unvarying, force is sufficient.

At impact on the edge of the floes, its shearing off occurred repeatedly along the line of least resistance. The floes crawled up by 1.2 - 0.8 m, wherein the area of individual chunks reached 1.5 - 2 m^2. Cases were observed where the floe, which had first crept up, slid down the support until it (the floe) went under water, moving under a new floe which had arrived during this time; after this, both floes advanced onto the support, breaking up together.

At encounter with particularly large floes (ice fields) occurrences are possible which are similar to the operation of ice aprons at movements of ice, as was noted earlier.

For example, on 26 April 1933, a floe was observed with a length of 130 m, having occupied about 2/3 of the river's breadth The floe moved obliquely in relation to the current. At impact with one of the supports, it turned and was carried against 5 bridge supports, which cut into it, having left smooth furrows in the floe. However, the floe did not break up.

4. Function of Bridge Supports During Period of Dispersed Ice

Frequently, when the ice is scattered, the small floes, having struck against an abutment, sustain local disruption at the point of impact. They then turn around the impact point and move away to the adjoining span, not being broken up by the ice apron, if other floes do not interfere. The departure of the floes away from the ice apron promotes the formation of backwater in front of the support. However, in case of the motion of larger floes, their splitting occurs similarly to the destruction of a large floe during the full-scale ice passage.

It is necessary to point out that the movement of more or less large floes in the period of scattered ice is quite possible under the conditions of ice passages in the rivers of Siberia, occurring during frequent ice jams.

In designing and building the supports of bridges and hydraulic engineering structures, it is also important to know not only the force of ice impact but also the point of its application. The pressure developing during the movements of ice is maximal but usually can not be the calculated value, owing to the low levels of motions, in connection with which the stability of a support, absorbing impacts in the lower part, will be greater than at other points.

However, the thickness of the ice pack under the Siberian conditions can be fairly great even during complete ice passage. In connection with this, we postulate that the most advantageous calculation conditions for the operation of supports are developed in one of the following situations:

a) either during destruction of a large floe during total ice passage occurring for many rivers of Siberia at high level;

b) or at encounter with a large floe of sufficient strength during the period of dispersed ice and at high water level. What has been indicated pertains only to the Siberian rivers.

At other conditions, with a different nature in the rising water (maximum level does not coincide with the ice passage), with local features precluding counting on the appearance of large floes during the high water level (as also takes place for many rivers in the European sector of the USSR), the pressure developing during the motion of ice can also be a calculated value.

5. Operating Features of Individual Piers of Overflow Weirs

The present study is devoted mainly to the question of the ice's action against the supports of bridges and hydraulic engineering structures; in connection with this, everything stated above also pertains fully to the individual piers located on the crest of the spillway dams. However, there are certain features which we must discuss.

The conditions of the breakup of an ice cover are somewhat complicated in the given case by the presence of the fall-off (decay) curve, developing in front of the spillway. Therefore during the approach to the spillway crest, the ice cover is broken up and this is aided by the operation of the piers. As early as the observations made by Prof. V.T. Bovin [74] on the passage of ice through the dam at the Volkhov Hydraulic Station, it was established that during the approach of an ice field, a crack is formed, oriented along the dam's axis and situated at a distance of 3 - 4 H. In its turn, the broken-off part of the floe divides into two parts above the spillway crest, while during descent over the spillway surface of the dam, it becomes broken further into 3 - 4 pieces, each of which is not larger than 3 - 5 m. Fig. 5 illustrates the conditions of the breakup of an ice field in the given case. In general, our observations confirm the findings of V.T. Bovin.

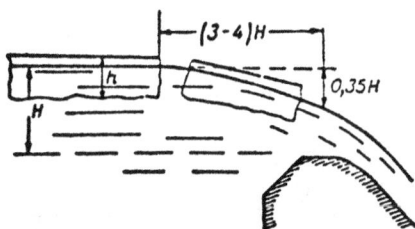

Fig. 5. Disruption of Ice on Fall-Off Curve.

The stopping of ice fields at the dam piers is possible but it should be borne in mind that cases of jamming on the crest have not been observed in the practice of operating dams. For example, on 2 April at the span of one of the dams, a floe stopped with a size of 15 X 10 X 0.35 m, but under the effect of other floes, it was quickly cut through by the abutment.

Thus, the piers of spillway dams are under somewhat better conditions as compared with the bridge supports. In this case, it is most likely to have a splitting of the floes (with a vertical cutting edge) or shearing--in case of an inclined ice apron (starling). Disruption from bending is less likely in connection with the abnormally greater development of the spring thawing processes, accompanied by a significant decrease in the ultimate shearing strength (of ice) along the crystals' axis.

It is therefore recommended to impart to the pier's cutting edge a tapered form in plane in conformity with the findings made in Chapter 23. In most cases, the application of an inclined cutting edge is superfluous since the disruption of the ice cover even without it (the edge) is greatly facilitated by the   decay curve.   Let us point out that at one of the hydroelectric power plants in the Soviet Union, we are utilizing piers with an excessively steep ice apron (1: 1/7), which naturally does not make sense for the reasons which we review in detail below.

Hence, in planning the individual piers and spillway dams, one should take into account a reduction in ice strength by 25 - 30% and an alleviation of the breakup of the ice cover, caused by the   decay curve.   It is obvious that the most probable at this time is the breakup of the ice cover by shearing or splitting.

6. Dynamic and Static Effect of Ice on Long Structures

a) Dynamic Action. The edge of a large ice field, encountering a dam, an ice-protective wall, shore bank, cofferdam or dock structure along its route, begins to break up, at this time exerting pressure of varying degrees.

With a sloped edge of the structure, the disruption occurs by way of breaking into segments, the length of which does not exceed 3 - 6 thicknesses of the ice. Similar cases have been specifically described by S.A. Katkova [75] for the Tsimlyanskaya HES (Hydroelectric Station), by A. I. Komarovskiy [1, 43] for many regions of the world, and by V.M. Samochkin [76] for the Yenisey River and are illustrated in Fig. 6.

During the climbing of an ice floe onto the sloped edge of a structure, it is possible for the floe to form hummocks and piles of very considerable height (up to 10 - 30 m), permitting one to estimate the amount of vertical pressure developing at this time.

With a vertical edge of the structure, it is possible to have the disruption of the ice edge by means of fragmentation with simultaneous hummocking, developing from the pressing out

of the ice by the advancing field. Let us point out that with
an inclined edge, the extent of the ice's action can not be particularly great (as was indicated below), and in certain cases
such pileups are even desirable.

Fig. 6.   Breakup of Ice Floes During Climbing
Onto Slope

The accumulations of floes, usually originating at the
beginning of the ice passage, then absorb the pressure of the
newly arriving ice fields, thus reliably protecting the structure. Such a pattern was observed by us at the cofferdams of
the HES under construction in Siberia.

With a vertical edge, the action can be more significant
and can lead to deformations or even the destruction of the
structure. Such cases have been described in the next chapter.

b) Static Effect. As is known, the temperature of the
lower surface of the ice is constant and equals 0°C; in connection with this, temperature fluctuations occur in the upper part
of the ice cover. Therefore, at fluctuations in air temperature
and hence in the ice cover, its bending and deformation occur.

In the large basins, thermal cracks often form; their
width can reach considerable dimensions, as is found e.g. in
Lake Baikal. According to data furnished by the Baikal Limno-
logical Station [77], the entire depth of the denuded ice cover

is usually intersected by many fine, and also wider, dry wedge-shaped cracks. A similar pattern was also observed by G.Ya. Kazub on the Ob' River [78]. At a temperature drop of only 3°, the reduction in the linear dimensions of the ice cover over the length of Lake Baikal comprises up to 120 m and it inevitably must break up. As a result, the entire ice cover of the lake divides into separate fields of 10 - 30 km in width. At warming, the ice cover expands and hummocks develop at the edges of the "main fissures", i.e. the huge temperature cracks. In the spring, the edges of the main fissures break up and displacements occur, accompanied by tremendous accumulations of ice.

In this way, there develops a very unique pattern in the deformation of the ice pack, complicated still further by the nature of the shoreline. In the case of sheer rocky shores (or of vertical edges of a structure), considerable forces of interaction develop, capable of causing certain amounts of damage. In the presence of gently sloping shores (or the sloped edges of a structure), we can observe the climbing up of the ice sheet according to the pattern described above. At abrupt temperature changes, ridges can form on the surface of the ice sheet.

Without doubt, the theoretical calculation of the value of pressure developing at this time becomes quite complicated and we can estimate only the order of this value's magnitude.

7. Abrasive Effect of Ice Pack

The abrasive effect of the ice is occasioned by the frictional forces originating between the moving floes and the shore or structure. In connection with this, damages develop in the form of furrows at the level of the ice passage. On certain rivers of Siberia, furrows are formed even on the rocky cliffs alongshore [40].

The above-water wall of one of the embankments was destroyed in 3 years owing to the appearance of deep furrows (1.5 X 0.3 m) under the abrasive action of the ice cover [82].

The shore revetments (especially the brushwood ones) and also the wooden (both the pile and the crib type) structures sustain intensive damage. Cases are on record when during a continuous autumn ice passage, with their sharp edges, ice floes have cut through piles in 5-6 hours [79]. Undoubtedly the frictional forces developing are linked with the value of standard pressure, the nature of processing the structure's surface, and the extent of its solidity.

8. Effect of Form of Abutment on Conditions of Its Operation

A. Effect of Inclination Angle of Ice-Cutting Edges to the Horizon

The observations described above pertain to the supports with inclined and vertical ice-cutting edges. Considering the function of supports with inclined edges, we conclude that their main purpose is to facilitate the breakup of the ice sheet. An inclined ice apron so to speak undercuts the ice field, thereby facilitating its more complete destruction. The sloped edge forces the floe to break up from fracturing and shearing rather than from crushing (as takes place with the vertical starlings) and thereby obstructs the possibility of the formation of ice jams, so likely under the Siberian conditions. Icebreakers climbing onto an ice sheet and crushing it under their weight operate in a similar manner. There can be no doubt that if the icebreaking vessels had vertical edges at the bow section of the ship, their attempts to provide the breakthrough of continuous icefields many kilometers in length would be doomed to failure.

B. Effect of Form of an Ice Apron in Plane

As was shown by the observations made by G.P. Perederiy [80], G. Karlsen and H. Streletskiy [68], the author [54], and also A.A. Trufanov [67], the effect of the ice apron's shape in plane is quite significant.

According to observations made by G. Karlsen and N. Streletskiy, the ice aprons of the hipped type **with width expanding** downstream functions poorly. Between the ice aprons, a jamming of the floes occurred which led (in 1944) to the washing away of 4 ice aprons even with very weak ice. The application of ice aprons of constant width proved more successful. In the ice passage in 1945, no damage was sustained by any starlings of this type. The same conclusion was reached by A.A. Trufanov and G.N. Petrov, who conducted laboratory studies of the question.

The purpose of the laboratory investigations was to find the most feasible form of an ice apron in plane and to establish an efficient mutual arrangement on the basic and outer ice aprons at the wooden bridges. A test was made of 23 different systems, wherein the evaluation criterion was provided by the number and nature of the impacts of ice floes on the supports, the presence or absence of ice jams, and an estimation of the trajectories of the floes' movement.

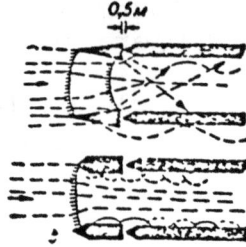

Fig. 7.   Paths of Movement of Floes with Use of
Starlings of Various Shapes

The modelling was conducted according to Eisner and Froude,
where, in the conversion to full-scale, we assumed: density of ice
passage--40% of the water surface; current velocity ranging from
0.5 to 3.0 m/sec; spans of bridge--10 m; thickness of abutment--
2 m; widths of starling--1.5, 2.0 and 3.0 m; inclination of cutting edge--1:2; and distances between abutment and starling--0.5,
3, and 8 m.

The results of the laboratory studies showed that the conditions of ice passage depend greatly on various local factors
(irregularity in distribution of velocities, impacts of floes
on one another), in connection with which the tests could fur-
nish only a qualitative description of the phenomenon.

The most interesting patterns of the arrangements of ice
jams and of the floes' trajectories are presented in Fig. 7. As
is obvious, it proved more efficient to use an ice apron which
was rectangular in plane. In this connection, ice jams are
possible only in the forward part and hence are separated from
the bridge supports. The triangular starlings cause the formation
of jams chiefly in the stern part of the ice aprons, which in
connection with this becomes exposed to the lateral pressure of
the ice.

At any form and width of the starlings, the increase in
the distance between them and the support greatly deteriorated
the conditions of streamline flow, and led to the formation of
local eddies and ice jams, as well as to the formation of a
number of frontal impacts of ice against the supports. Tests
of systems with various distances between the outer and basic
ice aprons demonstrated the feasibility of setting this distance
at 30 m.

At a reduction in the distance of 15 m, the floes lost
velocity owing to impact with the outer starling, then approached
the main starlings with slow speeds, which led to ice jams.

At an increase in the distance to 50 m, the ice floes changed their direction of movement, acquired during travel past the advanced post ice aprons, which also rendered the passage of ice more difficult. Meeting an inclined edge, the floe started to crumble from underneath and, advancing gradually onto the ice apron, became bent resiliently and broke off along the line of contact with the edge.

As a result, G.N. Petrov recommends: a) the application of rectangular ice aprons with a width equalling that of the support (abutment) and with an inclined edge. The axes of the ice aprons and the support should be parallel to the direction of flow; b) the superposing of the ice apron with the supports; and c) distance between the advanced post and basic ice aprons should be at least equal to the tripled dimension of the span.

It should be noted that basically G.N. Petrov has properly explained the operating conditions of the ice aprons at wooden bridges. However, the given experiments are inadequate to make possible a recommendation of any given quantitative characteristics (slope of ice apron's cutting edge, distance between advanced post and basic ice aprons, etc.), which should be indicated on the basis of more thorough investigations.

9. Brief Conclusions

Summing up, we can make certain general conclusions.

1. In the period preceding the opening of a river, one finds a certain decrease in the thickness of ice cover both owing to the thawing of the upper ice layers and also owing to the "eating away" of the lower floe surface by the warming water. As a result of this, there is a considerable increase in the roughness of the ice pack's lower surface.

2. The increase of water discharge into a river leads to a separation of the ice pack from the shores and to its floating up. The increasing current velocities in combination with the increasing roughness of the lower ice surface leads to the breaking away of the large ice fields, to their movement, and to displacements of the ice.

3. The thawing process begins with a weakening of the bond between individual ice crystallites; as a result, there is an intensive weakening in the ice's shearing strength along the crystal's axes. The flexural strength of the ice also decreases, but to a lesser degree.

Table 1

Typical Cases of Ice Destruction

Dimensions of floes	Extent of river surface coverage by ice	Nature of floe's disintegration	
		with sloped ice apron	with straight ice apron
1. Large ice field	total	At slow velocities (to 1.0 m/sec) it is cut through without splitting. The ice breaks from shearing or (if stronger) from bending in direct proximity to the pier	At slow velocities (to 1 m/sec), field is cut without splitting. The ice breaks owing to plastic compression and crushing. Stopping of floes is possible
	drift ice	At high velocities and small field dimensions, splitting of floe is possible--total or partial	At high velocities and smaller dimensions of field, with sharp outline of ice apron in plane, floe's splitting (total or partial) is possible
2. Medium	complete, scattered ice	Splitting of floe	Splitting of floe or its stopping, with spreading into adjoining span
3. Small	complete	Pier functions as a wedge, moving the floes aside. Floes jammed against pier become split	The floes stop, are turned by the current and move away to the adjoining span
	drift ice	Floes having struck against a pier, stop, are turned by the current and move off to the adjoining span	

4. The process of the breakdown of the ice sheet begins from local deformations in the contact area of the ice and the abutment. The force of interaction increases (in the initial period of the introduction of the support into the ice) and can attain values sufficient to stop the ice at the support, for a complete or partial splitting of the floe, or for its fragmentation.

5. After the introduction of the support into the ice, along its entire width, pressure on the support reaches a maximal value, not increasing later.

6. Typical cases of the breakdown of an ice pack are reflected in Table 1.

7. The close contact of an ice pack with a support is possible only with extremely weak ice which is not dangerous per se. With ice of any appreciable strength, the width of the support section contacting with the floe does not exceed 0.5 - 0.8 of the support width.

8. Pressure of the floe on the support varies through time; in this connection, the vibration of the support occurs.

9. The tilted edge of the support's ice-shearing edge facilitates the destruction of the ice sheet. In this instance, the ice mantle is broken from shearing or rupture and not from the crushing of the ice field, as does take place in case of the vertical starlings.

10. The most disadvantageous aspect of the supports' functioning under the ice motion conditions in the Siberian rivers is the encounter of a large floe with a support during full-scale ice passage or even during dispersed ice. Under other ice passage conditions (the noncoincidence of the highest levels of water with ice passage, low strength of ice), the most disadvantageous factor is the movement of ice.

11. In the planning of wooden ice aprons, their rectangular form with constant width is preferred. The use of ice aprons in plane facilitates the formation of ice jams in their stern part.

Chapter III

DISRUPTION AND DAMAGE TO STRUCTURE, CAUSED BY ICE PACK

1. General Concepts

The formation, accretion and movement of an ice pack in basins exposed to harsh climatic conditions frequently lead to various types of damages to bridges and hydraulic engineering structures and in isolated cases are accompanied by their total destruction.

In the period of autumn ice passage, the motion of ice floes which are thin but have sharp edges can lead to damage to shore revetments (especially the brushwood type) and also to wooden structures (individual piles, crib works, light pile structures, trestle designs of embankments etc.).

After the formation of a stable ice cover, at abrupt variations in air temperature, a temperature expansion of the ice occurs, accompanied by an increase in pressure on the adjoining edges of dams and of other engineering structures.

In individual cases, the pressure of the ice sheet during its expansion can lead to appreciable damages (deflection of the structures from the vertical, damage to sluice gates, shifting of individual parts) or even to the destruction of the structures if effective measures are not adopted to cope with the developing pressure.

It is evident that the static pressure of the ice mantle will have special significance for structures of great extent (dams), existing under harsh climatic conditions and can be taken into account for individually standing supports most often only in case of the possibility of the development of unilateral pressure of ice or pressure at various levels. However, we have observed damages caused by static ice pressure to separately standing structures.

During the spring excess of water or during fluctuations in its level in the winter period, we have recurrently noted damage and destruction to the structural parts frozen into the ice. There occurred a pulling out of piles, damage to the shore-reinforcing structures, manifested in the turning of rocks in bridges (sometimes over a very considerable area), disruption of brushwood rows, the outer part of brushwood reinforcements etc. We also observed the twisting out of individual stones from rock masonry in the body of the structures.

During the spring ice passage, the ice motion can be accompanied by significant damage to the shores, structures, moreover, this can be particularly serious at formation and breaching of ice jams.

It should be indicated that even in the absence of large ice jams, the amount of ice pressure originating at impact of large floes drifting at great speed can be quite significant.

2. Certain Causes of Damage to Structures and Accidents Caused by Ice Pressure

An analysis of the data presented in the literature and also personal observations of the ice passage conditions on the Siberian rivers permit us to list and analyze certain examples of the breakdown of structures and to review the damages caused by ice pressure.

Cases of Dynamic Pressure of Ice

1. On the St. Lawrence River (U.S.) in winter, the ice sheet imprisoned the stone bridge abutments, which at opening of the river were broken up; one of the fragments was moved 11 km from its previous location [42,43,8].

2. On 25 January 1789, owing to the formation of an ice jam during ice passage, three abutments of the de Tour Bridge were washed out on the Loire River (France), and in addition four arches (with spans of 24 and 36 m) were destroyed [80].

3. By impacts of floes during the ice passage in 1852, one of the old spillway crib dams on the Muskingum River in Ohio was destroyed, in which the dam crest was wrecked for a width of 15 m [7].

4. By impacts of floes, there were repeated damages (1857, 1865, 1873, 1875) to the Columbia crib dam in Pennsylvania. The length of the damaged sector ranged from 270 to 1200 m [1, 5].

5. In 1882, the crib spillway dam on the Monongahela River in West Virginia [1] was damaged for a length of 37 m.

6. In 1893, an ice debacle in the U.S. destroyed the dam of the Mewan Waterworks [1, 43].

7. In 1897, an unfinished dam at St. Antoine Falls on the Mississippi River [1, 43] was partly destroyed.

8. In Idaho, the impact from a huge ice field 0.40 m thick knocked from its vertical position (in an upper point) a ferroconcrete floodgate tower 26 m in height and 6.6 m in diameter [1].

9. By the impact from an ice block frozen to the river bottom and raised during flood stage, there was pushed into the downstream water a metal gate at the Keokuk hydroelectric dam; the gate weighed 13 tons and the force of the impact was estimated at 92 tons [1].

10. In 1910, during ice passage, the bridge across Swatora-Creek River near Hummelstown [1, 81] was destroyed. In 1912, the dam with a height of 6.4 m was destroyed by ice pressure. The ice thickness was 0.9 m [60].

11. In 1915, on certain rivers in the northwestern part of our country, the spring ice passage carried away almost all the newly built wooden bridges resting on crib-type supports [81].

12. The crumbling of a steel bridge with a span of 100 m built across the Connecticut R. at Brattleboro was caused by a catastrophic flood during which the ice pressure (ice with a thickness of around 1.0 m) toppled a girder weighing 270 tons and with a length of 100 m. Calculations show that the ice pressure reached 140 tons/m^2 [81, 85]. It should be emphasized that the wrecking of the given bridge in 1920 was not the first time, since even prior to that it had been carried off 7 times by the ice [6].

13. The dynamic pressure exerted by the ice on 21 March 1926 caused serious damage to the pier of Big Rock Bridge below Franklin (Pa.); as a result, 2 girders were knocked into the river [81, 6]. During this same flood stage, at Reno Bridge, a span structure was shifted by 0.30 m [6].

14. In spring of 1929, as a result of an ice jam on the Yenisey R. at the Ladeyskiye Sandbars, in several minutes the ice broke down the wharf structures. The dredger "Sib XV" was ripped from its moorings and the backwater dam was wrecked. The force of the ice jam which had disrupted cut off an island on the Ladeyskiye Sandbars and completely altered its configuration.

15. In March 1936, a bridge across the St. Joanna R. (Canada) was wrecked during an unprecedented flood caused by extremely heavy rains, at simultaneous jamming of the ice below the bridge crossing. The pressure of the ice damaged almost all the piers and overturned the span structures [7].

16. The ice jam having formed on 27 Jan. 1938 in the source of Niagara R. wrecked the span structure of an arched metal bridge near a railroad, destroyed the steamship wharfs and buildings and also almost completely flooded the generator room of the Ontario hydroelectric plant on the Canadian side of the river. The ice jam was caused by the fact that part of the ice pack on Lake Erie had been drifted by the wind into Niagara R. which was still ice-covered. The water level rose 15 m. After two shoves, the arched bridge with a span of 256 m was wrecked, and the ice broke through a window into the hydroelectric station, covering the generator with a layer up to 5 m deep. Drag lines were used to remove the ice [7].

17. In spring of 1944, **on** one river in the European part of the USSR, 4 ice aprons of a wooden bridge with spans of 10 m were carried off by an ice sheet of only 0.25 m thickness. The destruction was caused by the jamming of ice in the stern parts of the tent-style ice aprons [68].

18. In the spring of 1948, ice passage in Siberia damaged the crib pier of the docks.

19. During World War II, damage occurred to several bridge abutments of one of the large railway bridges across a waterway of great width, described by A.S. Ol'mezov and G.S. Shpiro [69]. Let us examine this case in more detail.

The bridge abutments consisted of high ferroconcrete grating resting on 20 metal piles filled with concrete (8 of them at an angle). By the onset of the ice passage, the starlings (planned for erection) had not yet been built. During the entire winter, the water at the abutments remained open and their deformations were caused by an ice field transported by the wind and current. Initially, the ice field pressed against the eastern shore and then approached the bridge and started to press against the supports. The ice field deflected, from the vertical position, the piles of several supports by 0.5 - 1.0 m; after this, they were restored by a jolt to their original position, breaking the ice.

The ice field, moving at a speed of 0.3 - 0.4 m/sec, at encounter with the supports, began to be cut through by them at a rate of 0.01 - 0.02 m/sec, wherein strips of open water remained beyond the supports. At the supports, the ice thickness comprised 45, 50, and 60 cm respectively, while at 100 m above the bridge, the depths were 60, 100, 30, 27 and 22 cm. Having traversed 50 m, the ice field stopped, having formed a continuous ice jam above the bridge, and this was favored by the cool weather which had set in.

In the area of one of the supports, the ice accumulation reached the bottom (at basin depth of 5 - 8 m). Blasting operations produced no effect. At ensuing displacements, many supports were greatly damaged. The ferroconcrete gratings proved to have been shifted: one had been moved aside by 8 m and downward by 2 m, a second was moved aside by 2.6 m, downward by 0.2 m, and a third had been turned by 45°, in which one of its edges had been lowered by 3.5 m and another of them by 2.9 m. Almost all the metal piles were bent by 10 - 15°; some were broken off.

A.S. Ol'mezov and G.S. Shpiro [69] theorize that one of the reasons for the wrecking was the unsatisfactory division of the bridge opening into spans, since with short spans, the relationship between the vertical and horizontal forces acting on the abutments were unfavorable. In addition, the lack of ice aprons and an excessively steep slope of the forward piles (7:1) were reflected. The ice pressure of an individual support was estimated at 270 tons.

20. I.N. Shafir and R.I. Ginsberg in their report [82] describe interesting cases of damage inflicted on marine hydraulic engineering facilities in the USSR ports.

The above-water wall of one of the embankments was wrecked in 3 years after construction owing to the appearance of deep furrows (1.5 m X 0.3 m) from the influence of the ice pack.

At the breakwater of one of the ports, there were ripped from the structural surfaces rocks (including chunks) which had been frozen to the ice sheet, and these rocks were carried out to sea.

The pier of one of the ports with a length of 427 m consisted of cribs 10.2 X 7 X 8 m, loaded with rock and installed 8.3 m apart. The spaces between the cribworks were filled with a pile pier, having been formed of 24 piles each with a diameter of 26 cm. On 6 May 1941, an ice field with a width of around 10 km and a thickness of 0.6 m was brought to the pier by the current and tide. The field was moving at great speed, forming some wave breakers ahead of itself. After the impact and the ensuing ice pressure, the pier sustained serious damage: 15 cribs were displaced (some up to 8 m distance) and damaged.

The authors describing the damage and examining the deformation of the piers, estimated the force of impact needed for shifting each crib, by the value

$$P = Q \cdot f = 328 \cdot 0.3 = 100 \text{ tons,}$$

which yields 14.3 tons for 1 running m of the crib and 23.8 tons for 1 m^2 of contact area of the ice and the pier. Here $Q = 328 =$ weight of crib in water, and $f = 0.3$ --the coefficient of the crib's friction in blue clay.

21. In December 1955, in one of the large rivers in China, a solid bridge support 4.1 m in diameter was shifted. Damage occurred under the effect of a large ice field 0.4 m thick propelled by the wind, at air temperature of 20°C. A. M. Ryabukho [86] estimated the possible pressure force at 300 tons.

In addition to the cases tabulated, there are numerous local damages to bridge supports, having specifically been observed in many Siberian rivers and also damages to hydraulic engineering structures, e.g. of embankments and docking facilities (shearing of iron mooring posts on the Syas R. in 1881) and also accidents to ships. There are even more numerous examples of damage and losses of seagoing vessels. Thus, in

1913, 1931, 1934, and 1938, the vessels Karluk, Chukotka, Chelyuskin, and Rabochiy were beset and crushed by ice; the steamer Rabochiy was crushed within 12 minutes [38].

Cases of Damage During Static Ice Pressure

Records also show the wrecking or damaging of structures caused by static ice pressure; certain of these cases are presented below.

1. In February 1899, at variation in temperature from -22° to -8°C, damage by ice 1.2 m thick was observed; this was at a dam in Minneapolis, having been shifted from the vertical by 25-30 cm. Somewhat later, a part of the dam slid down over an extent of 52 m along a crack and collapsed [60, 43]. Calculations indicated that the pressure amounted to 16.7 tons/m^2.

2. A dam across the Saranac R. at Morissonville, U.S. on 15 Jan. 1912 was partly damaged by ice, when part of its crest with a length of 24.5 m and a height of 1.25 m proved to have been displaced [60,43].

3. In winter of 1902, ice moved the baffle rocks, in a dam, each with a volume of 0.67 m^{3}m and which had been placed on dowels [1, 60].

4. The pier at one of the bridges in Canada weighing around 1,000 tons was subjected to unilateral pressure of an ice sheet 0.305 m thick and it moved 5 cm from the vertical position. After ice was cleared around it, the pier assumed a vertical position; calculations showed that the ice pressure reached 71.3 tons/m^2 [1].

5. The rock abutments of a bridge on a pile foundation across Kent Bay in Belleville, Ontario, under the effect of ice sheet pressure, shifted from the vertical by 5 - 30 cm and returned to their upright position only after the cutting of the ice. Apparently the occurrence just described is ascribable to the fact that the piles driven into the soft muddy ground became deformed under the effect of considerable ice pressure [1, 81].

6. After the spring of 1926, the tower of one of the lighthouses shifted by 1°30', while in March 1927, it had shifted by 2°08' [83].

7. The flat gates of the dam at the Keokuk Hydroelectric Station on the Mississippi R. with a span of 9.75 m in the winter of 1916-1917 became warped by 6.2 cm, corresponding to a pressure of 12 tons/running meter [80], or 21.5 tons/m^2 [1].

Summing up briefly, we can establish that the maximum number of damages and accidents having been attributed to the effect of the ice pack, falls in the period of spring ice passage and testifies to the tremendous destructive force of ice at this time. During the formation of ice jams and after their breakup, the destructive action of the ice is particularly great; this must be kept in mind during the planning of structures.

In actual isolated instances, the static pressure of ice also attained a significant level, however only in the absence of cutting around the supports etc., and at unilateral pressure

SECOND PART

SURVEY OF PHYSICO-MECHANICAL PROPERTIES OF RIVER ICE

Chapter IV

ACCRETION AND DISAPPEARANCE OF ICE COVER

1. Initial conditions

The formation of ice on rivers, lakes and reservoirs occurs as a result of the processes of heat exchange between the water and the atmosphere. The basic factors determining the intensity of the accretion and disappearance of the ice sheet are the meteorological conditions, and the hydraulic regime of the basin (current velocity, depth, mixing conditions of the water masses).

In a basin with slight water mobility (lake, reservoir, pond), at limited mixing, the cooling of water masses is nevertheless accompanied by their mixing. The surface layers chilled to 4°C being heavier, sink into the near-bottom areas, whence the warmer and lighter water masses rise.

Thermal convection continues until the entire water mass has a temperature of 4°C. Then convective exchange stops and a further drop in water temperature occurs, especially on the surface of the basin. As soon as the surface temperature reaches 0°C, the crystallization processes begin, leading to the appearance of a stable ice cover.

In a river flow, the ice formation processes are complicated by turbulent mixing which causes a more uniform cooling of the water and a later freezing over of the waterway.

The lack of an ice cover under conditions of low air temperatures and adequate current velocity provide the necessary conditions (supercooling of water, intensive heat exchange with its surface diversion of latent heat of ice formation, presence of crystallization nuclei, etc.) for the formation of ice within the water, developing in the form of friable spongy masses covering the bottom of the river and also in the form of suspended ice crystals.

In this way, the freezing of flowing water is accompanied by the formation not only of surface but also of submerged intra-water ice both on the river bottom and in a suspended state

The upwelled masses of sludge (frazil ice) then freeze together and float down river, forming the autumn ice passage.

A further decline in air temperatures causes the freezing together of individual floes and formation of stable ice, after which the formation of ice within the water ceases. In the presence of open river sectors, it can be formed and be borne by the current to the sectors located farther downstream.

Under the effect of warming and the flood caused by it, from the wind or the breakthrough of water trapped under the snow, the ice sheet which has formed can be broken up; as a result, a repeated autumn ice passage will develop. With the arrival of heavy frosts, the ice passage stops and finally yields to a stable ice sheet.

In the straight reaches of a river with a slow current, with freezing during calm frosty weather unaccompanied by thaws, an ice cover is formed having a smooth surface. At freezing over of a river in a period of unsettled weather, with recurrent ice passages, in the sectors with an irregular pattern of current, an ice sheet is formed with a rough surface, covered with hummocks, at times attaining considerable height.

2. Certain Features of Ice Structure

The various conditions of the formation of an ice cover permit us to differentiate:

1. Water (crystalline) ice. It is formed during direct freezing over of a surface layer of clear water without admixtures of other previously formed types of ice. As a result of the effect of the autumn ice passage, wind and current, the upper layers of this ice have an irregular finely-crystalline structure. Since the further increment in the ice thickness occurs under calmer conditions, the underlying quite transparent layers of greenish or light bluish hue have a regular columnar structure.

In the ice sheet of all, even the very slightly mineralized basins, the ice crystals are separated from each other by the finest interbeddings of mother solutions of salts having been released during freezing. It is conventional to assume [12] that owing to their lower melting temperature, the thawing starts from these interstratifications.

In this connection, as a result of the difference in the volumes of ice and melt water near the interbeddings, areas develop with reduced pressure. The forming vacuum, although it is slight, is sufficient for the penetration of air into the air column; this greatly increases the total number of cavities, makes the ice turbid and porous, breaking into individual crystal lites with a sectional dimension of 1 - 4 cm, divided by streaks (veinlets) with a diameter ranging from 0.1 to 1.0 mm. In connection with this, the anisotropy of the ice sheet in the spring period becomes expressed much more distinctly and establishes the dependence of the ice's mechanical properties on the orientation of the pressure.

We sometimes witness the breakup of a floe into individual needles of snow-white hue, of much smaller diameter (1 - 5 mm) and height (10 - 30 cm), also oriented vertically. Often among a regular crystalline structure, we can find an inclusion of air bubbles, forming during the rapid freezing owing to the separation of gases dissolved in the water and of intra-water ice in the form of irregular turbid, fluffy formations.

2. Snow ice is formed from snow soaked with water and then frozen together, when such snow has fallen on the surface of the water ice.

3. Water-sludge ice, forming during the freezing of water containing sludge is less transparent than the water ice and has a more irregular structure.

The inclusions of sludge are represented in the form of turbid spots of irregular shape. It should be noted that in nature, we encounter an extremely diversified combination of ice of various structures; nevertheless, we can indicate that in most instances, the upper layers of the ice pack have a snow and water-sludge structure, while the basic layer of the ice pack consists of ice having formed from water.

3. Dynamics of Variation in Depth of Ice Pack

Depth of the ice pack during the stable ice period changes constantly under the effect of the actual conditions--pattern of air temperature, depth and compactness of snow accumulations, and velocity conditions of current.

In the equations utilized in practical calculations [87], thickness of ice pack h_{π} is ordinarily linked with the sum of negative mean diurnal air temperatures $\sum_t {}^{\circ}$ for the considered

period of time and sometimes with the depth of snow layer, h_{cH}, on the ice pack.

F.I. Bydin's [88] equation has become most popular:

$$h_\Pi = 2 \sqrt{\Sigma t} \quad [\text{cm}] \tag{4.1}$$

or

$$h_\Pi = 3.67 \sqrt{\Sigma t + 6h_\Pi^2} - 9 h_\Pi \tag{4.2}$$

In the absence of snow on the ice, V.V. Piotrovich [89] recommends that we assume the thickness of ice

$$\text{from } h_\Pi = (\Sigma t)^{0.695} \tag{4.3}$$

$$\text{to } h_\Pi = 1.25 \, (\Sigma t)^{0.62} \tag{4.4}$$

This first formula is given for clear, windy weather, while the second is for overcast skies and slight winds.

B.A. Apollov [90] recommends the formula

$$h_\Pi = 1.8 \left[1 + \frac{1}{h_{cH}} \sqrt{\Sigma t} \right], \tag{4.5}$$

in which h_Π and h_{cH} are adopted in centimeters.

In connection with the complexity of the problem, it is of undoubted interest to consider the observations of the actual pattern of increase in ice thickness. The Laboratory of Ice Thermics of the Siberian Branch at the USSR Academy of Sciences organized the studies, the results of which are explained in the article by I.P. Butyagin [87].

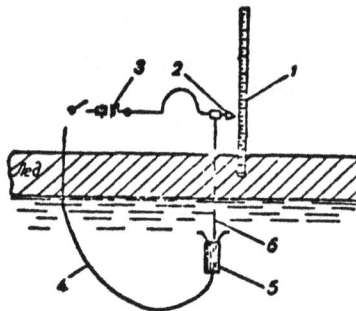

Fig. 8. Diagram Showing Electric Ice Measuring Device: 1 - measuring rod; 2 - indicator; 3 - battery; 4, 6 - conductors; and 5 - weight with supports.

They can be described briefly as follows.

The conventional procedure in ice-measuring observations in use by the hydrometeorological service does not provide a sufficiently accurate description of the process of variation in the thickness of river ice. The results of the periodic measurements in the holes and during drilling are distorted greatly by the local, often appreciable differences in thickness of the ice cover. In addition, the presence of permanent holes disrupts the natural conditions of ice formation. For improving the quality of periodic ice-measuring observations, use was made of special electric- measuring units designed in line with the principle suggested in 1933 by P.I. Syrnikov [91]; their arrangement is given in Fig. 8. The inclusion, within the system, of a portable battery, causes the heating of a nickel conductor, on which under the ice cover, a weight is suspended with supports in the upper part. Heating the wire permits the weight to be raised freely, establishing according to the stationary rod the variation in depth of ice cover from the lower surface. The accretion and thawing of the ice from the upper surface is noted directly from the rod. The rod for the readings is fastened rigidly on two supports frozen deeply into the ice cover.

The application of such devices permits us to conduct periodic ice-measurement observations at strictly appointed locations, without disturbing the natural conditions of the ice's accretion and melting.

Analyzing the given observations, I.P. Butyagın [87] concludes:

The formation of the ice cover on the Ob' R. is determined chiefly by the temperature regime of the ice-over period. Satisfactory results in computing the ice thickness on the Ob' are provided by F.I. Bydin's formula. Special ice-measuring observations conducted by a hydrologic expedition from the Transport Power-Engineering Institute of the Siberian Branch in the USSR Academy of Sciences permit us to refine the indicated relationship for the conditions of the river's upper flow.

Variations in thickness of the ice pack occur variously from its lower and upper surfaces. From the lower surface, the ice accretes continuously, most intensively in the first 2 months of the ice-over. Later, the intensity of ice formation diminishes perceptibly, gradually abating up to the beginning of the spring warming trend. From the upper surface, the increment of ice has the nature of brief but abrupt increases in its thicknesses, associated with the icing formations. The process of ice thawing during the spring period develops continuously with increasing intensity. Thickness of the ice cover in spring decreases by 20 - 50%.

By spring, the lower surface of the ice field is always covered with blister-like depressions with a diameter of 15 - 20 cm and a depth of 5 - 10 cm.

In addition, the fluctuations in current velocity according to width of river and the irregular depth of the snow cover cause a varying depth of the ice sheet, which leads to its increased roughness. Our measurements of the depth of the ice cover on the Ob' R. for an area of 250,000 m^2, conducted prior to the opening of the river, yielded fluctuations ranging from 0.16 to 1.25 m at an average thickness of 0.73 m based on data from 224 measurements. In this connection, the limits in the fluctuations in ice thickness at each of the 11 sections comprised from 0.20 to 0.81 m. In this manner, we should take into account the considerable roughness of the ice up to the period of ice-out.

4. Features of Ice Cover on Reservoirs

Undoubtedly, the slower current speeds, higher water temperature and great depth of water flow are inevitably reflected on the structure, thickness and strength of the ice cover of the reservoirs.

The observations conducted by the Transport Power-Engineering Institute of the Siberian Branch of the USSR Academy of Sciences in the large reservoirs and lakes of Siberia permit us to conclude that in this case:

1. The ice conditions in supported downstream (below dam) waters differ significantly from the natural ones in connection with the abrupt increase in depths, decrease in current velocities, variation in the thermal regime of the basin, and increased role of the wind.

2. Thickness of the ice cover increases somewhat (by about 15-20%) in connection with the earlier freezing and the slower current velocities.

3. The clearing of the large reservoirs of ice occurs 10 - 12 days later than on a river in a natural state.

4, Strength of an ice cover and its depth decrease significantly (up to 60-70% of the former values).

5. Structure of the ice cover is mainly crystalline.

Chapter V

ELASTIC AND PLASTIC PROPERTIES OF ICE

1. Basic Tendencies in Deformation

Just as any other physical body, ice deforms under a load,
wherein we can distinguish elastic and plastic deformations.
Even in the case of the simplest stressed state, the amount of
the ice's deformation depends on a number of factors (for example,
the amount of stresses, effective time of load, temperature of
ice and its structure), which cause a brittle or viscous plastic breakdown.

It is common knowledge that under the effect of sufficiently extended load, ice can alter its form without appreciable indications of breakdown, manifesting the property of
fluidity. The movement of glaciers and also the Hess experiments
[116] at outflow of an ice stream from the opening of a vessel
illustrate this. However, under other conditions (low temperatures, dynamic nature of loads), the ice has a typically brittle
breakdown.

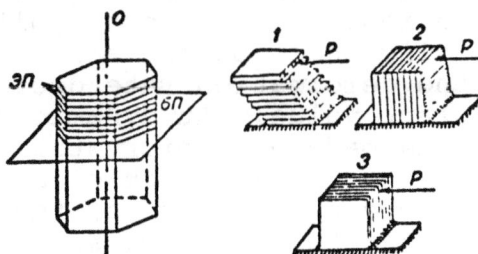

Fig. 9. Pattern of Deformation of Ice Crystal
According to K.F. Voytkovskiy: o = optical axis of
crystal; δn = base plane; $\mathfrak{z}$n = elementary plates;
and P = shearing force.

Let us describe, chiefly following K.F. Voytkovskiy [94],
the main tendencies in the deformation of monocrystalline and
polycrystalline ice. An ice crystal can be represented as an
aggregation of many thin plates, perpendicular to its (crystal's)
optical axis, capable of moving relatively easily in relation to
one another.

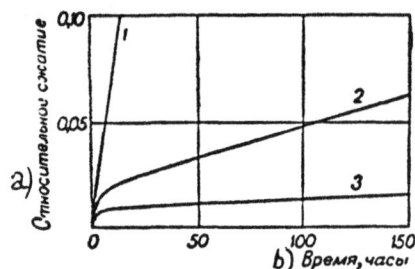

Fig. 10. Stress-Strain Curves of Specimens of Semi-crystalline Ice According to Glenn's Data:
1 - σ = 6.1 Kg/cm^2, t = -0.2°C; 2 - σ = 6.1 Kg/cm^2, t = -6.7°C; 3 - σ = 6.0 Kg/cm^2, t = -12.7°C. Key: a) Relative compression; and b) Time, hours.

We have shown in Fig. 9 a diagram of the deformation of an ice monocrystal during plastic (1) and elastic-plastic (2) deformation. In case 3, the plates after slight elastic deformations can bend owing to the loss in stability and relative displacements, leading (at an increase in forces) to the breakdown of the crystal. In this way, in the monocrystal, it is possible to have both the elastic deformations of the plates and also their relative plastic displacements.

The deformations of semicrystalline ice are more complex During compression, after a slight elastic deformation, plastic deformation begins; creep is expressed with a constant rate (at slight stresses) or with increasing rate of deformation (at stresses of the order of 10 Kg/cm^2 and higher). Figure 10 illustrates these tendencies. At a smooth application of the bending forces, initially there occurs a rapid increase in deformations and then the deformation rate decreases, tending toward a certain constant value. At dynamic application of the flexural forces, columns made of semicrystalline ice are easily broken down. The breakdown begins with cleavage in the area of the neutral axis, followed by fracture of the ice in the distended zone and drifting into the compressed zone [24].

The semicrystalline ice is thus deformed under the effect of the following closely interrelated factors: elastic and plastic deformations of crystals, their mutual displacements and finally, the breakdown of the most overstressed crystals. The processes of recrystallization and unique regeneration of the ice's structure are proceeding simultaneously.

-52-

The nature of deformation is determined by which of the processes (disruption of the bonds or their restoration) is prevalent.

Thus, the tendencies in ice deformation are quite complex and as yet insufficiently studied. Therefore, in an estimation of the possible amount of ice pressure on structures, it is still necessary to rely chiefly on the mass determinations of the mechanical properties of an ice cover under conditions simulating nature as nearly as possible. In this connection, we have succeeded, without delving into the details of the ice disruption mechanisms, in developing general concepts about the conditions of the structures' operation. Let us note in addition that the procedure for the determinations should be deliberated profoundly and thoroughly.

2. Elastic Properties

The elastic properties of ice have still been studied very insufficiently, since in connection with the smallness of ice's elastic limit (around 0.5 Kg/cm^2) under natural conditions and in the laboratory, one almost exclusively finds ice in a plastic state.

It would thus have seemed that the application of the conclusions of the elasticity theory to the problem of the action of ice on engineering structures is impossible and one should proceed from other assumptions. However, a number of researchers [14, 24, 112, 113] have established that under known conditions (low ice temperatures, slight stresses, fixed range of deformation rate), the dependence between the deformations and the stresses is close to a linear one, and in this manner the possibility of utilizing Hooke's Law is formally substantiated.

Along with this, many authors [94, 114 etc.] have established that the plastic deformation of ice develops even under very slight stresses and that in proportion to a decrease in the loading rate, the dependence between the deformations and stresses will deviate more and more from a linear one.

Thus, in an analysis of stressed state, it is necessary to take the time factor and the viscous, plastic deformations of ice into consideration. Let us present certain data on the characteristics of ice's elastic properties.

The elastic modulus (E) is usually found by two methods. The static method is based on measurements of the amount of deformation during tests of ice for compression, expansion or bending. It should be noted that it is quite inadmissible to

·measure the elastic constants far beyond the limits of elasticity. As a result of this, many determinations of the elastic modulus by the static method are more exactly uniquely characteristic of ice plasticity than elastic constants [115].

The dynamic method is based on the measurements of rate of propagation, in an ice cover, of elastic fluctuations, and undoubtedly deserves preference over the static method.

A determination of the elastic modulus E was conducted by many authors, starting with Bevan (1824) and having continued until recent times. A more or less detailed explanation of the question and a summary of the results derived therefrom are given in the reports by B.P. Veynberg [12], K.F. Voytkovskiy [94], Ye. M. Lin'kov [97], V.P. Berdennikov [26] and by the author [53]. Certain of the findings of interest for the problem formulated by us have been included in Table 2.

Table 2

Elastic Modulus E of Ice, Based on Data of Determinations

Researcher	Year	Type of Ice	Method	Conditions of loading	$-t^\circ C$	$E, t/cm^2$	Biblio.
MacRae	1885	lake	static	bending	1-3	41-72	[12]
					5-7	59-104	
Pinegin,V.N.	1923	river	"	"	5-9	12	[14]
					15-19	21	
				compres.	5	48-84	
Shul'man, A.R., Ivanov,K.Ye., Kobeko,P.P.	1946	"	"	bending of ice field	44	44	[94]
Kartashkin,B.D.	1943-45	"	"	expansion	6-8	37-50	[24]
				bending	2-21	35-62	
Voytkovskiy,K.F.	1954-58	artif.	"	"	1-4	25-65	[94]
Butyagin,I.P.	1957-58	river	"	"	3-9	30-40	[31]
Erwing, Crary & Thorne	1934	lake artif.	dynamic	artif.	5-15	88-98	[98]
Nakoya	1958	glacial	"	"	9	90	[94]
Bordennikov,V.P.	1948	artif.	"	"	2-40	88-97	[26]
Boyle & Sproule	1931	lake	"	"	10	94	[104]
					35	109	

As we shall observe, the results of determining the elastic compliance coefficient (modulus) are quite contradictory, especially those obtained by the static method. Analyzing them, we come to the following conclusions.

1. For ice, the elastic modulus is a value which is difficult to determine owing to the low value of the elastic limit.

2. The most reliable method of determination is the dynamic one, permitting a reduction in the inevitable errors and providing more coordinated results ($E \approx 90$ tons/cm^2).

3. At a more or less prolonged loading, of interest are the observations made by Ivanov, Shul'man and Kobeko [94], as well as by Butyagin [31], obtained during tests of ice fields for bending. The value obtained for the elastic modulus ($E \approx 30$ 44 tons/cm^2) is somewhat tentative but it can be utilized in estimating the value of static loads acting on a structure for an extended period.

4. With a drop in temperature, the modulus increases slightly. We can agree with V.P. Berdennikov [26] regarding the possibility of disregarding the temperature fluctuations of the ice.

5. With an increase in stress, the "tentative" elastic modulus decreases perceptibly, as was demonstrated by tests conducted by V.N, Pinegin [14] and K.F. Voytkovskiy [94].

Let us point out that this can be explained in terms of the increasing role of ice creep. The data obtained in this connection doubtless cannot be considered as elastic constants or even as a deformation modulus.

The shear modulus (G) is usually found by testing ice samples for torsion since thereby conditions of pure displacement are created.

A collation of the findings, compiled by B.P. Veynberg [12] and K.F. Voytkovskiy [94] permits us to present certain of them, representing the maximum interest for the problem which we have formulated (Table 3).

As we shall see, quite analogous to what has been said above, the actual G-value should be found by the dynamic method and should be estimated at 25-35 t/cm^2.

Table 3

Shear Modulus G of Ice, Based on Data of Determinations

Researcher	Year	Type of ice	Method	Load conditions	-t° C	G, t/cm^2	Literature
Veynberg	1905	river	static	torsion	0	10	[12]
Koch	1914	lake	"	"	5	16	[117]
Brockamp & Mothes	1930	glacial	dynamic	seismic	0	28-30 25	[118]
Erwing, Crary & Thorne	1934	artif.	"	tors, fluct.	5-15	34	[98]
Kartashkin	1943-45	river	static	tors.	10-16	13-18	[24]
Voytkovskiy	1958	artif,	"	"	4	12-18	[94]

The provisional value of the modulus for estimating the values of static loads can be assumed to equal 20 - 15 t/cm^2.

The Poisson coefficient (μ) for isotropic materials at elastic deformations is linked with the elastic modulus E and shear modulus G by the relationship:

$$\mu = E/2G - 1, \tag{5.1}$$

utilizing which, we find the possible fluctuations

from $\mu = 90/2.30 - 1 = 0.50$ to $\mu = 40/2.15 - 1 = 0.33$.

Direct measurements of the Poisson coefficient for the ice on Tom' R. were made by V.N. Pinegin [99], having recorded its increase with an increased load. Let us cite the available recommendations for choosing the calculating value μ :

V.P. Veynberg . . .μ = 0.36 $\pm$ 0.13 [12]

B.D.Kartashkin . . μ = 0.34 (for temperatures from −5° to −16°C [24]

B.A. Savel'yev .. μ = 0.36 [100]

It is obvious that at an increase in load and hence at a more rapid manifestation of the plastic deformations (transpiring without variation in volume at $\mu = 0.50$), the Poisson coefficient can acquire higher values.

Moreover, since ice is not an elastic body, in the process of the increase in pressure and at a rise in temperature, the plastic deformations become more and more significant and in general the concept of the Poisson coefficient loses physical meaning. N.N. Petrunichev [104] proceeding from a review of elastic-viscous deformations, derived equations linking the stresses and the rate of relative deformation. This provides the possibility of computing the variable in respect to time, the apparent value for μ. Thus, at tensile stresses of 20 Kg/cm^2 and coefficient of viscosity $8 \cdot 10^{-7}$ Kg·hr/m^2, the following values of the Poisson coefficient have been obtained:

t =	0	100	500	1,000	2,000	3,000	seconds
μ =	0.30	0.31	0.31	0.35	0.36	0.41	

A similar nature in the increase in the apparent Poisson coefficient was indicated in the experiments made by V.N. Pinegin [14]. Let us also note the undoubted effect of the force's orientation relative to the crystallization axis.

In conclusion, we wish to point out that the elastic properties of ice have still been quite inadequately studied; many determinations of the elastic constants have been made based on unsatisfactory procedures, and therefore fail to reflect the actual pattern of the phenomena.

3. Plastic Properties

For determining the amount of ice pressure on structures, of maximum interest are the relationships between the plastic deformations, effective stresses, duration of load effect and ice temperature. Referring those interested to the reports by B.P. Veynberg [12, 103], K.F. Voytkovskiy [94], V.V. Lavrov [36, 101, 102], S.S. Vyalov [35], A.R. Shul'man [25, 115] and N.N. Petrunichev [104], we will limit ourselves to a discussion of only the main questions necessary for estimating the value of ice pressure on structures.

Relationship Between Deformations and Stresses. In studying the ultimate compressive strength of river ice [19], 568 diagrams were taken, linking the increasing force with deformation of the sample being compressed. An analysis of the diagram permits us to establish the following features for them.

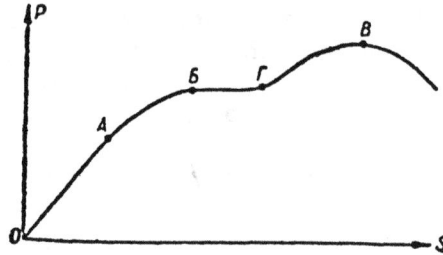

Fig. 11. Relationship Between Compressive Force P
and Sample Deformation S.

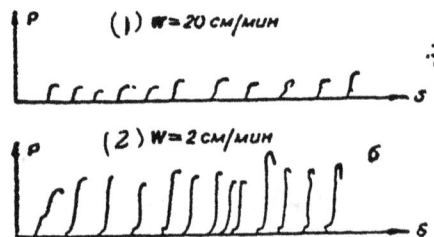

Fig. 12. Nature of Compression Diagrams at Variation
of Deformation: 1 - at S = 0.0033 sec^{-1}; and
2 - at S = 0.0167 sec^{-1}. Key: 1) W = 20 cm/min;
and 2) W = 2 cm/min.

Initially, the force increases quickly (Fig. 11), usually
approximately according to a linear law (sector 0A of the diagram), and we then observe a certain increase in the deformations (Аб). Subsequently, it is possible to have an even more
rapid increase in the deformations (бГ), and finally, after a
certain hardening, disruption of the sample occurs (point B
in the graph).

A similar pattern was derived in testing cubical ice
samples for uniaxial compression, with an edge dimension of
10 - 12 cm, at deformation rate of 0.0017 - 0.0033 sec^{-1} and
ice temperature having varied in the limits from 0° to -15°C.
Direction of the force was perpendicular to the crystallization
axis; the samples were cut from the natural ice cover on the
river.

At an increase in the deformation rate by 10-20 times
(up to 0.0417 sec^{-1}), the nature of the diagram changed somewhat. Deformations increased and became complicated owing

to the appearance of local fissures, while the value of the destructive force decreased. The graphs depicting both cases are reflected in Fig. 12.

Diagrams similar to this type were obtained by N.A. Tsytovich and M.I. Sumgin [119] and also by A.R. Shul'man [115], I.P. Butyagin [30] and other authors. Royen [63] on the basis of experimental research, suggested the empirical formula:

$$\varepsilon = \frac{\Delta l}{l} = \frac{B_1 \sigma}{R - \sigma},$$ (5.2)

where σ, R = effective and disruptive stress; ε = relative deformation of compression; and B_1 = a constant value.

Equation (5.2) can scarcely be considered as having physical meaning, since it includes the R-value, not comprising a constant, and typifying the material's deformation.

Relationship of Deformations to Temperature. The effect of temperature on the value and nature of ice deformations is indisputable. In Royen's opinion, relative deformation is linked with ice temperature by the dependence:

$$\varepsilon = \frac{B_2}{1 + \theta},$$ (5.3)

where θ = ice temperature in °C, without the minus sign; and B_2 = a constant, depending on dimensions and shape of sample, duration and value of the load.

As we shall observe, the maximum plastic deformations occur at temperatures close to the melting temperature of ice. However, the actual nature of the relationship of ice's plastic properties with the temperature is more complex. Temperature influences not only the ε-value but also the nature of the dependence (5.3). The results of Glen's tests [120] illustrate this (see Fig. 10) (Glen studied the deformations of compressed specimens of polycrystalline ice at practically constant stress (6.0 - 6.7 Kg/cm^2) and various temperatures (ranging from 0° to -12.7°).

The tests by K.F. Voytkovskiy [94] on the bending of ice beams and the twisting of pipes showed that a variation in the value of creep rate at assigned conditions of deformation is described fairly well (in the temperature range from -1° to -40°C) by the empir

$$\gamma_\theta = \frac{1 + \theta_0}{1 + \theta} \gamma_0,$$

(5.4)

where γ_o = experimentally established rate of steady creep at θ_o; and $\dot{\gamma}_\theta$ = rate of steady creep at any temperature θ.

According to the author's assertion, this equation is applicable for various forms of deformation (compression, bending, frequent shearing) at steady creep rates and not depending on the temperature distribution of the internal stresses.

Of interest are the laboratory investigations permitting us to develop the dependence of the stresses originating in an ice sheet at varying rate of increase in its temperature. G.E. Monfore has reported on the tests performed in the U.S. Bureau of Reclamation [123]. In a thermally insulated chamber, uniform temperature distribution was provided of an ice sample held between the soft jaws of a press; the samples dimensions were kept unchanged. The experiments showed that with an increase in the rate of temperature rise, an increase in pressure occurs, initially quite rapid; an unambiguous relationship between the temperature rise and the pressure increase was lacking; the values characterizing the maximum pressure at an increase in temperature by 9°C per hour are listed in Table 4.

Table 4

Dependence of Stresses on Ice Temperature

Initial t° in °C	Max. pressure, Kg/cm^2	Initial t° in °C	Max. pressure, Kg/cm^2
- 5	6	-20	16
-10	10	-30	21
-15	13	-35	23

Dependence of Deformation on Time. The question of most importance for us and at the same time still inadequately researched is the one concerning the dependence of deformation on time.

Currently, in the solution of a number of engineering problems, particularly for determining the static pressure of ice on structures, at times use is still made of the Royen formula, having the form:

$$\varepsilon = \frac{C\sigma\sqrt[3]{t}}{1+T},$$

(5.5)

where t = times of load's effect in hours; T = absolute value of negative ice temperature; σ = stresses, Kg/c,^{2}; and C = a constant having the value of (60 - 90)10^{-5}.

It should be noted that from the experiments with ice, Royen obtained only the dependence of relative deformation on temperature (5.3), more or less agreeing with later determinations (e.g. 5.4). However the dependence of ε on stress σ and time t of its effect was derived from tests with paraffin and cannot be considered reliable. Thus Eq. (5.5) indicates the damping of deformations with time and the linear relationship of deformation and effective loads, which was not confirmed by experience or more detailed studies. N.N. Petrunichev [104] presents data from the experiments of V.P. Berdennikov [122], indicating a very complex and unique relationship of the rate of relative deformation and stresses. Figure 13 illustrates the results of V.P. Berdennikov's determinations.

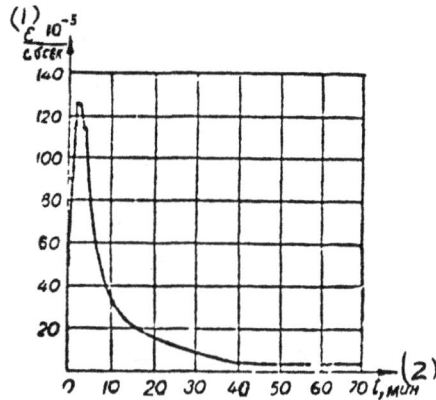

Fig. 13. Curve Indicating Rate of Relative Deformation in Relationship to Time (after V.P. Berdennikov) Key: 1) $\varepsilon/\Delta t$, sec; and 2) t, min.

Often the plastic deformation of ice is regarded as a flow of viscous liquid in which the rate of deformation is determined by the viscosity factor (η). However, as is indicated by B.V. Proskuryakov [121], the η-value depends not only on temperature but also on the effective time of the load, type of deformation and orientation of ice crystals. Reliable determinations of η are lacking since the values derived differ by thousands of times from each other. Usually the viscosity factor for ice was computed proceeding from the assumption that ice complies with Newton's Law for viscosity, i.e. a linear relationship was assumed between the stress value and the deformation rate.

In the reports by Glen [120], K.F. Voytkovskiy [94], Haefeli and Steinemann [124,125], it was shown that there is no linear dependence between the stress value and the deformation rate, and that the η-value is not a fixed physical constant but a variable depending on many factors.

Summing up what has been said, according to a number of studies, it is necessary to make the following conclusions:

1. At the modern level of knowledge, one cannot simply interpret the flow of ice with the aid of any given model (viscous liquid, elastic-viscous body etc.).

2. The flow of ice is a process which depends significantly on time. Initially, during a short time interval, one finds a nonstationary process, then alternating with stationary flow. In certain cases, it is subsequently possible to have an increase in the deformation rate, often associated with a change in the sample's geometry.

3. The calculation formula of the Royen type is too rough an approximation of reality, since it provides an incorrect concept of the process of ice flow attenuating through time.

4. The undertaking of careful experiments in the laboratory and under natural conditions is necessary, based on a special procedure, with ice and uniform structure.

Chapter VI

ULTIMATE STRENGTH OF ICE

1. State of the Question

In connection with the great practical value of information concerning the strength of ice for the planning of
structures, ice crossings, airfields, the setting up of ice
arches etc., this strength has been studied by various researchers and especially intensively from the end of the 19th Century
up to the present time.

An outstanding role in the study of the mechanical properties of ice was played by the Russian and Soviet scientists,
having performed comprehensive original and extensive research
into the question. In 1897, B.P. Vasenko [10] studied the ultimate strength of ice on the Neva R. during crushing, breaking
and bending, and in 1900, S.O. Makarov [11], while planning the
Yermak icebreaker, also conducted studies of ice strength during
crushing.

A systematic study of the strength of ice on the Tom R.
was undertaken in Siberia by B.P. Veynberg [13], Ye. Bessonov
[105] and V.N. Pinegin [14]. The latter tested more than 200
samples of ice from the Tom' R. for compression, tension, bending,
and also established the elastic constants for ice. In 1921,
B.N. Sergeyev [15] studied the mechanical properties of ice on
the Volga R., while from 1923-1928, reports were published by
Yu. A. Pedder [106] and V.I. Arnol'd-Alyab'yev [107], having
studied the ice on the Angara R. and the Gulf of Finland.

From 1934-1936, detailed studies of the strength of ice
on the Svir' R. prior to ice passage were conducted by the
Scientific-Research Institute of Hydraulic Engineering in Leningrad [16] on 297 samples, as well as by V.S. Nazarov [17],
F.F. Vitman and N.P. Shendrikov [18] on ice in the Kara Sea.

During the same years (1934-1938), extensive laboratory
studies were undertaken on the mechanical properties of ice in
the Ob' R.; more than 700 samples were tested. The findings
clarified the question of the effect of the ice structure, dimensions and form of a specimen, orientation of pressure relative
to freezing plane, temperature and rate of loading on the
ultimate compressive and flexural strength of ice. For the
ensuing years, we should mention the studies made by B.G.Korenev,

S.M. Izyumov [23], A.S. Ol'mezov and G.S. Shpiro [69], and also the interesting report by B.D. Kartashkin [24], having studied the elastic and plastic properties of ice and having established its ultimate compressive, bending, torsional and tensile strength

In 1939, I.P. Troshchinskiy [12] proposed an original method for determining the flexural strength of ice under natural conditions (by the method of "keys", which was then utilized by Yu. N. Neronov [28]). The studies by I.P. Troshchinskiy laid the foundations for the tests of the strength of large ice masses occurring under natural conditions, which is quite significant.

Interesting diversified research in accordance with an extensive program was conducted by the Laboratory of Ice Thermics of the Siberian Branch at the USSR Academy of Sciences (I.P. But-yagin) under natural conditions [30,32].

In recent times, A.R. Shul'man [25] and V.P. Berdennikov [26] investigated the resistance of ice to plastic deformations and determined its elastic constants. We should also make specific reference to the major works by Prof. N.N. Zubov [27], B.P. Veynberg [12], K.F. Voytkovskiy [94],also especially dedi-cated to the questions of the mechanical properties of ice.

However, in spite of the large number of studies, the question of the actual strength of ice, especially at the time of river breakup, cannot yet be considered adequately investi-gated for the following concepts.

1. The strength of the ice pack up to the time of the spring ice passage usually decreases under the effect of solar radiation and a rise in the temperature of air and water. At the same time, the experiments on studying the mechanical properties of ice were usually conducted during the winter on specimens taken from relatively permanent ice and do not reflect its actual strength during the spring. Attempts to determine the strength of ice during the spring ice passage are still few in number.

2. Under actual conditions of the operation of structural supports, one observes a movement of the ice fields with a velocity up to 1.0 - 3 m/sec which, as our investigations [19] have shown, should be reflected perceptibly on the value of the ultimate crushing strength of ice. At the same time, almost all the studies of which we are aware were conducted under condi-tions of static load.

3. The investigations indicate the extreme dissimilarity of individual ice samples, in connection with which it is necessary to undertake large-scale tests which would permit the obtainment of reliable rather than random results. At the same time, only a few large-scale tests have been conducted. V.N. Pinegin [14] tested 200 samples, M.M. Basin [16]--297, the authors -- 1012 samples [19], I.P. Butyagin [30] conducted 1014 tests, while most of the remaining researchers limited themselves to a small number of tests (up to 10), which explains in part the extreme hit-or-miss nature of the results derived.

4. Most of the researchers worked on cubes or bars of ice, having had the possibility of becoming deformed in all directions. However, under actual conditions of the operations of supports, cases occur when the disrupted part of a floe does not have the possibility of being freely deformed owing to the influence of adjoining sectors of the ice sheet.

5. Then under conditions of ice passage, the ice usually has a temperature close to 0°C and is in a melting stage. For quite understandable reasons, a study of the mechanical properties of ice during this period becomes difficult, in connection with which we lacked reliable information specifically for these temperatures.

6. Finally, let us note that the cutting edge of a support usually has a fairly complex form: it was therefore necessary to conduct additional studies on determining the resistance of ice to the penetration of variously sized punches into it.

Taking into account what has been stated, from 1934-1961 the author, and then I.P. Butyagin, conducted extensive investigations on the study of:

a) the degree of reduction in strength of the ice cover up to the time of the spring ice passage;

b) the effect of the ultimate strength of ice during crushing and bending, rate of loading, phenomenon of local grinding, ice temperature, orientation of pressure relative to the freezing plane, and dimensions of the sample; and

c) resistance of ice to the penetration of variously shaped punches into it.

Employing the available data, we can explain the value of the ultimate strength of ice to compression, bending, shearing and stretching under the conditions of spring ice passage.

In connection with this, let us limit ourselves to the utilization of data approaching the conditions of spring ice passage and therefore most interesting for solving the problem formulated.

2. Ultimate Strength of Spring Ice During Compression

Effect of Orientation of Pressure Relative to the Freezing Plane. Numerous studies have shown that in the orientation of pressure along the axis of the crystals (perpendicularly to the freezing surface), the ultimate strength at compression ($R_{\parallel}$) is always greater than at any other direction of pressure.

Let us present a compilation of certain findings of the ultimate compressive strength at ice temperatures close to 0° (Table 5).

Table 5

Ultimate Strength of Ice at Uniaxial Compression

Researcher	Year	Nature of ice	$t.°C$	$R_{\perp}$	$R_{\parallel}$	$\dfrac{R_{\parallel}}{R_{\perp}}$	No. of samples
V.N. Pinegin	1922	river	0 – 2	28	36	1.28	74
K.N. Korzhavin	1934	"	0–0.6	11	30	2.72	44
M.M. Basin	1934	"	0	12	16	1.33	187
Ice Station	1930	"	-3.0	19	41	2.16	72
Bruns	1934	"	0	--	--	2.00	25
K.N. Korzhavin	1938	"	0	15	—	--	28
			-3.6	27	—	--	24
B.P. Veynberg	1940	"	-3.0	combined data		1.28	717
F.I. Bydin		"	0	10	—	—	17
K.N. Korzhavin	1956	"	0	10	—	--	22
P.V. Fedorov	1941	"	-1.0	19	29	1.51	8
I.P. Butyagin	1953–57	"	0	12	—	--	116

As we shall observe, the effect of the orientation of pressure relative to the axis of the crystals is variable and increases in the process of spring thawing of ice.

B.P. Veynberg's data ($R_{||}/R_{\perp} = 1.28$) obtained after a generalization of the results of many experiments conducted at various temperatures thus characterizes the values of $R_{||}/R_{\perp}$ for the specimens not subjected to the spring thawing processes. The same should be noted relative to M.M. Basin's experiments, having worked with ice which was still stable.

Our experiments (having been conducted from April - May on samples having been prepared from floes tossed onto shore by the ice passage) reflect a more significant decrease in ice strength owing to the weakening of the bond between the crystallites and for the ratio $R_{||}/R_{\perp}$ yield the value of 2.7.

With a decrease in ice temperature during the winter, the ultimate compressive strength of ice increases. Detailed studies [19] permit us to recommend the determination of the ultimate compressive strength of river ice (at temperature ranging from 0°C to -12°C) according to the equation:

$$R_{\perp} = 11 + 3.5 \ Kg/cm^2, \qquad (6.1)$$

where t = absolute value of negative ice temperature.

Thus we conclude:

1. A natural ice cover has considerable anisotropy; its effect is abruptly intensified with the development of spring thawing processes.

2. The resistance of ice to the forces directed along the axis of crystals ($R_{||}$) is always greater than the resistance to the forces directly perpendicularly to the crystallization axis ($R_{\perp}$).

Table 6

Value of Ratio $R_{||}/R_{\perp}$

Nature of ice	Temperature	$R_{||}/R_{\perp}$
Ice cover without appreciable signs of weakening of bond between the crystals..........	from 0 to -3°	1.3 - 1.5
Ice cover with appreciable weakening of bond between crystals.....................	0°	2.0-2.7

3. We can assume the values of ratio $R_{||}/R_{\perp}$ shown in Table 6.

4. Ultimate strength of river ice at compression perpendicular to the crystals' axis at temperature of around 0° comprises from 10 to 12 Kg/cm^2.

5. For the lower temperatures, ice's ultimate compressive strength can be computed according to Eq. (6.1).

3. Ultimate Strength of Ice During Local Crushing

At the encounter of an ice field with bridge supports or a hydraulic engineering structure, the process of the floes' disruption begins, accompanied by the penetration of supports into the ice.

Since the supports usually have a sharp outline in plane, the width of the contact area gradually increases to a maximum value equalling the support's width. It is obvious that in this case, the ice cover experiences the phenomenon of local crushing

In the experiments usually being practiced for the crushing of cubical samples of ice, the latter can be freely deformed in all directions, whereas in case of a local nature of loading of the ice plate, its deformations become difficult. In this manner, the unloaded sectors of the floes play the part of a clamp, increasing the ice's ultimate strength. For a clarification of this question (which has still been studied little), special tests were undertaken which we have explained briefly below [22].

Test Conditions

For the preparation of samples, we utilized the ice from the Ob'R. at Novosibirsk, where specific attention was directed to the correctness of orientation of the principal dimensions of the specimens relative to the crystal's axis.

The strips of ice were trimmed with a specially designed device, comprised of a wooden box; over its open upper side, a wide plane tool could be slid. After this, the strips were cut into specimens of the required sizes and stored in an unheated chamber.

Desiring to exclude any attendant factors, we chose samples from clear ice without cracks, vesicles, bubbles or stray inclusions. Therefore under natural conditions, the ice should have somewhat less strength than that which was obtained in our studies.

On the day of the experiment, the samples in the laboratory were put in a thermostat-box   surrounded by thawing ice where they were kept for 3 - 4 hours, until the ice temperature had reached 0°. We selected such a temperature in order to bring the tests close to the actual conditions existing during the spring ice-out.

Fig. 14. Diagrams Reflecting Tests of Prismatic Samples of Ice: a - with upper lining; and b - with two linings.

Pressure was oriented perpendicularly to the crystals' axis, as takes place under actual conditions of the supports' operation. The approach rate of the support planes of the machine varied from 1.5 - 2 cm/min.

As a preliminary step, for obtaining data on the ultimate compressive strength of ice, (at temperature of 0°, perpendicular to the crystals' axis), we conducted the crushing of 22 samples measuring 70 X 70 X 70 mm. Then for revealing the effect of local crushing, we tested 177 blocks, some of which were loaded through solid metal strips with a width of 1.2 - 14.0 cm. The dimensions of the blocks varied from 70 X 70 X 200 to 70 X 70 X X 300 mm. The diagram of loading is evident from Fig. 14.

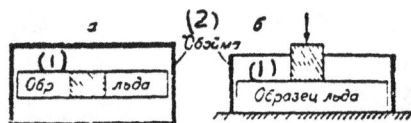

Fig. 15. Diagram Showing Tests of Ice Blocks in Band (Clamp): a - plane; and b - section. Key: 1) ice sample; and 2) clamp.

Desiring to explain the value of the ultimate crushing strength of ice under the conditions of the encounter of large ice fields with supports, in addition we tested 65 blocks enclosed in a rigid clamp as shown in Fig. 15. The freedom of deformations of such blocks was restricted only from the ends, while there were no obstructions for displacements in lateral directions. It was specifically under such conditions that the sector of large ice field edge, contacting with the support, was situated. As is known, in this connection the deformations of the loaded sector of the field upstream and across the river are made difficult, while both downward (toward the water) and upward, such deformations are quite possible. In this manner, undertaking tests on the blocks enclosed in the clamps, in the tests we attempted to simulate as nearly as possible the actual operating conditions of the ice aprons on the structural supports (ice temperature, orientation of pressure, nature of constraint imposed on deformations).

Type of Breakdown of Samples

Soon after the application of a load, a slight plastic deformation usually developed; owing to it, the transmitting pressures of the lining were lowered into the ice column by 1-2 mm.

Then, in the central part of the transparent ice samples, we detected traces of intercrystalline displacements in the form of web-like fissures, originating at the limits of individual ice crystallites. The number of such fissures gradually increased, propagating most often at an angle of 25 - 35° to the vertical (in the direction of the support areas).

Finally, two through cracks were formed and many of the ice crystallites became separated from one another. Certain ice crystallites became deformed and acquired a milky-white color, revealing areas of most intensive deformations. In a number of instances after the formation of cracks, one observed a peeling upward, and a bulging of the ice plates into the sample's upper edge (Fig. 16).

Fig. 16. Nature of Sample's Breakdown During Local Crushing.

Let us note that in the cases of uniaxial compression of blocks, their lateral parts were subjected very little to breakdown,while during biaxial compression, the entire specimen was cut through by a series of fissures. This nature of the breakdown of specimens testifies to the complex stressed state and to the decisive role of the shearing stresses. In our case, the deformation process was caused by intercrystalline displacements which during biaxial compression became difficult; this is explained chiefly by an increase in the ice's ultimate strength in this case. In this way, as a result of the tests, one should anticipate a significant effect, on the ice's ultimate strength, of the ratio of the crushing area's width to the block's length.

Results of Tests

Results obtained from tests of 199 samples are listed in Fig. 17 and can be interpreted by the following empirical relationships:

1. At any constant ratio b/h, the value R_{cM} is described by the equation:

$$R_{cM} = R_1 \sqrt[3]{\frac{B}{h}} \, , \qquad (6.2)$$

where R_1 = ultimate compressive strength at b/h = 1.

2. In its turn, the R_1-value can be computed according to the expression:

$$R_1 = R_o \sqrt[3]{\frac{h}{b}} \, , \qquad (6.3)$$

where R_o = ultimate strength during tests of ice cubes for compression.

Taking Eq. (6.2) and (6.3) into account, we find

$$R_{cM} = R_o \sqrt[3]{B/b} \qquad (6.4)$$

The dropping of sample thickness h from the final formula is normal, since, loading the sample over its entire thickness, in essence we were experimenting under conditions close to those existing in a plane problem. The validity of the findings

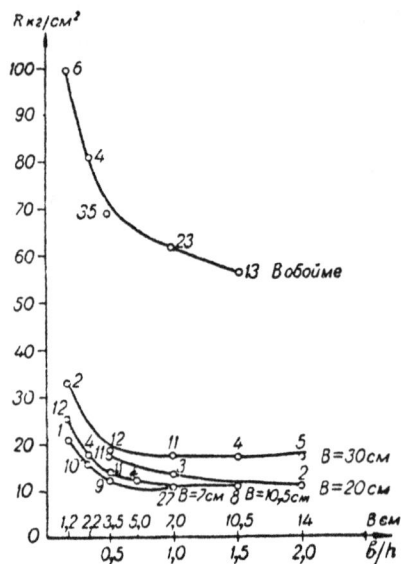

Fig. 17. Effect of Local Crushing on Ultimate Compressive Strength of Ice. Numbers at the curves show
number of tests; upper curve = tests of samples in
metal clamp. Key: a) R, Kg/cm^2; b) B, cm; b/h;
and c) In clamp.

indicated is confirmed by the comparison made of the experimental data vis-a-vis the computations. As is known, Eq. (6.4) is
also applied in the calculatinns of stone, concrete and ferroconcrete designs for local crushing. In this way, river ice in
respect to resistance to local crushing under the given conditions does not represent an exception among certain other
materials.

Let us examine the conditions of the transfer of the de-
pendences derived to the phenomena of ice breakdown by the supports under bridges and hydraulic engineering structures during
the spring ice passage.

The R_{cM}-Value Under Actual Springtime Ice Debacle Conditions

In a transfer to actual operating conditions, it is
possible to utilize Eq. (6.4).

In this connection, it is necessary to have the following
points in mind:

1. At introduction of a support into an ice field, a complex stressed state develops in it (the field). Directly at the support, there appears a zone of disrupted material, then being replaced by zones of the development of plastic and elastic deformations.

As is known from the theory of load distribution in linearly deformed masses, the stresses in the elastic zone attenuate fairly intensively (in proportion to distance from the point of forces' application) and finally, at a certain depth, they become so slight that their consideration lacks practical significance. Assume the width of the diagram of stresses still being considered by us constitutes B_1 (Fig. 18). Obviously, then the calculated floe width B_p can not be assumed to be greater than the value B_1, since a further increase in B no longer exerts any practical effect on the ice's ultimate compressive strength.

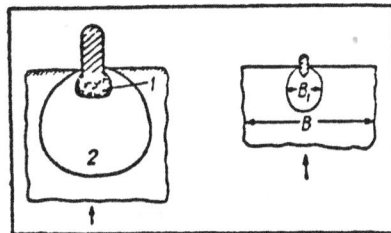

Fig. 18.   On Stress Distribution in an Ice Field:
1 - zone of plastic deformations; and 2 - zone of elastic deformations.

In addition, it is necessary to have in mind that the large ice fields, by the time of the spring ice passage, are usually interspersed with a system of fissures, dividing them into separate segments. Therefore in the initial period of disruption, the entire floe does not participate in the phenomenon, but only that part of it which the support encounters. To be sure, after the supports have penetrated into the ice, the adjoining segments are included in the process, but by this time, ratio B/b_{cM} has already decreased. These concepts also dictate the limitation of the floe's calculated width.

2. We experimented with blocks of relatively slight thickness, having rested on a practically absolutely rigid base, which of course was reflected on the test results. Under actual conditions, there would be a more uniform environment.

As is known, a rigid base creates a more concentrated distribution of stresses than in the case of the effect of a punch into a homogeneous medium. In connection with this, it is necessary to correct the value obtained.

3. Our tests, although conducted at a temperature of around 0°, were done on samples taken from "selected" ice (without cracks, inclusions, internal cavities), not yet subjected to intensive thawing. The bond between the crystallites was still fairly stable (strong). Therefore we should also correct the derived value of R_{cM}, taking into account the structural features of the ice pack during the spring ice-out.

Choice of Calculated Width of Floe

For finding the propagation limits of stresses (actually significant values) let us utilize the conclusions from the elasticity theory concerning the effect, on the edge of a linearly deformed half-space, of a load distribution on a strip with width b.

At the present time, this problem has been solved chiefly by N.P. Puzyrevskiy, N.M. Gersevanov, M.I. Gorbunov-Posadov, K. Ye. Yegorov, O. Ya. Shekhter, V.A. Gastev, V.G. Korotkin and other scientists.

In the given case, the application of the methods from the elasticity theory is justified since the stresses are distributed in a fairly homogeneous plate with slightly varying thickness, structure and elastic properties.

In order to validly operate with the formulas from the elasticity theory in application to the deformation of solid masses, as is known it suffices to know the linear relationship between the stresses and strains. Therefore the finding derived will naturally not be applicable to the zone directly contiguous to the support but can be used for explaining the distributions of stresses less than the proportionality limit.

From 1938-1950, we took 568 diagrams of compression of ice samples and 72 diagrams of the cores from punches driven into the ice. Analyzing them, we observe that the assumption about the linear deformability of ice at slight loads can be adopted. Taking into account that, according to the St. Venant principle, the distribution of stresses in an elastic half-space will not depend on the load distribution on the surface (starting from a depth roughly greater than 1.5 b), but is linked only with the value and the position of the resultant, we assume the pressure on the support to be evenly distributed.

In this case, the normal stresses (σ_x, σy) and tangents (τ) can be computed with the following equations:

$$\sigma_x = -\frac{p_0}{\pi}\left(\beta_1 + \frac{1}{2}\sin 2\beta_1 - \beta_2 - \frac{1}{2}\sin 2\beta_2\right);$$

$$\sigma_y = -\frac{p_0}{\pi}\left(\beta_1 - \frac{1}{2}\sin 2\beta_1 - \beta_2 + \frac{1}{2}\sin 2\beta_2\right);$$

$$\tau_0 = -\frac{p_0}{\pi}\left(\cos 2\beta_2 - \cos 2\beta_1\right)$$

(6.5)

(the values for p_0, β_1, β_2 are clear from Fig. 19).

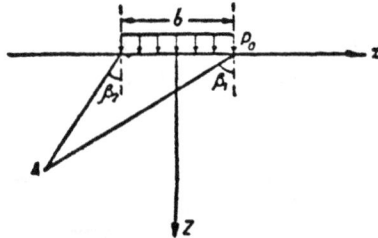

Fig. 19. On Determination of Stresses in Elastic Half-Space.

Utilizing these relationships, we expanded the pressure distribution curves in an elastic half-space, constructed by various authors. The derived data have been illustrated in Fig. 20. As we shall see, the area of propagation of normal stresses has the following width: at $\sigma = 0.05\, p_0$, $B_1 = 10\, b$; $\sigma = 0.03\, p_0$ $B_1 = 14\, b$; $\sigma = 0.02\, p_0$ $B_1 = 20\, b$. Taking into account that the disruptive compressive stresses during the spring ice passage usually do not exceed 5 - 8 Kg/cm^2, we will limit ourselves to considering the stresses less than 0.10 - 0.15 Kg/cm^2, which comprise around 2 - 3% of p_0.

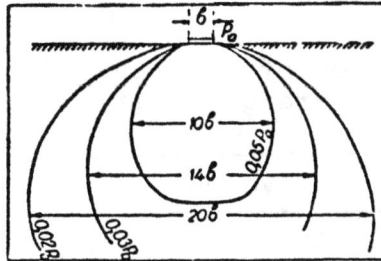

Fig. 20. On Selection of Computed Width of Floe.

In connection with this, let us limit the width of the calculated floe with the value:

$$B_1 = 15\, b.$$

Then, during the cutting through a large ice field by a support, the value of the ice's ultimate compressive strength can be calculated with Eq. (6.4), but cannot exceed the value:

$$R_{ck} = R_0 \sqrt[3]{\frac{15 b_0}{b_0}} = 2{,}47 R_0.$$

(6.6)

Rounding off, we can assume $R_{cM} = 2.50\ R_0$.

Refinement of R_{cM} -Value

As we have already indicated above, our experiments were conducted with blocks lying on the pads of a press which could be assumed as an absolutely rigid base.

The reports by K. Ye. Yegorov, N.N. Maslov, N.A. Tsytovich and other researchers established that the presence of a rigid base leads to an increase in the stress in the layer of material. In accordance with the theory of the functioning of a two-layered base, the increase in stresses reaches the values shown in Table 7.

In this manner, for practical purposes, the stresses in the contact plane exceed those in the elastic medium by 25-40%. Since the sum of compressive stresses in any horizontal section equals external force P, we can conclude that a rigid base promotes the concentration of stresses according to the pattern in Fig. 21.

There is hence no doubt that, if there had not been a rigid base, disruption would have taken place at greater forces, since the stresses would have been distributed more evenly.

Table 7

Nature of Distribution of Stresses with Biaxial Base
([Note]: Notations are given in Fig. 21).

$\frac{z}{H}$	Value $\frac{H}{b}$		
	0,5	1,0	2,5
0,00	1,000	1,000	1,000
0,20	1,020	1,020	1,020
0,40	1,051	1,045	1,040
0,60	1,085	1,105	1,000
0,80	1,163	1,315	1,195
1,00	1,246	1,382	1,440

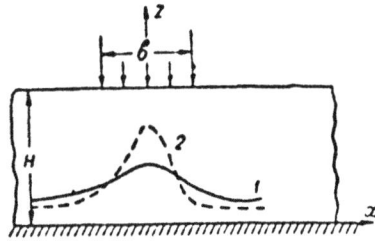

Fig. 21. Effect of Rigidity of Base: 1 - diagram of
stresses in homogeneous medium; and 2 - diagram of
stresses in two-layered medium.

Allowance for Weakening of Ice in Pre-Spring Period

As was shown above, the mechanical properties of river ice
are usually studied during the winter when the ice is still fairly
strong and sometimes even has a "choice" quality.

Therefore, although the tests were conducted at a temperature close to 0°, under actual conditions one should expect less
strength of the ice both in connection with the spring thawing
processes and owing to the greater heterogeneity of the spring
ice masses under natural conditions.

The studies by I.P. Troshchinskiy, Yu. N. Neronov and
I.P. Butyagin [12, 28, 30] confirm this conclusion fully. A particularly significant decline in strength (by 2-3 times) occurs
in the rivers flowing southward where the ice-out is preceded
by a prolonged warming. On the northern and Siberian rivers
where the opening is often caused by the mechanical effect of
flood water breaking up the ice sheet, the decrease in strength
will naturally be less. In this case, the extent of reduction
in ice strength can be assumed to equal 10 - 40%.

Calculated Value of Ultimate Compressive Strength

Taking into account what has been said, by the period of
ice-out, the ultimate compressive strength of river ice (perpendicular to the crystal's axis at static load) can be assumed
to equal:

a) for the rivers of the North and Siberia (flowing northward), in which the ice sheet does not succeed in being subjected
to appreciable weakening prior to the ice breakup,

from $\dfrac{11 \cdot 1.15}{1.10} = 12$ Kg/cm^2 to $\dfrac{11 \cdot 1.15}{1.40} = 9$ Kg/cm^2; and

b) for the lower reaches of rivers in the European part of the USSR (flowing southward), in which the river breakup is preceded by appreciable changes in the ice structure,

$$\text{from} \quad \frac{11 \cdot 1.15}{2.0} = 7 \text{ Kg/cm}^2 \quad \text{to} \quad \frac{11.0 \cdot 1.15}{3.0} = 5.0 \text{ Kg/cm}^2.$$

Brief Conclusions

1. During cutting through a large ice field by a support, the field is situated under conditions of local crushing, in connection with which the ice's ultimate strength increases to the value:

$$R_{CM} = R_0 \sqrt[3]{B/b_{CM}}.$$

2. The width of the calculated floe cannot exceed the support's width by more than 15 times, therefore the ice's ultimate strength is limited by the value:

$$R_{CM} = 2.50 R_{compres}.$$

3. At static load (to which the periods of ice's thrusts correspond), the value $R_{compres}$ can be assumed to fall within the limits:

$$R_{compres} = 6 - 10 \text{ Kg/cm}^2.$$

respectively for ice passages of medium and great force.

4. Effect of Deformation Rate on Value of River Ice's Ultimate Strength Original Conditions

Under the conditions of spring ice passage, the structural supports are subjected to the effect of the ice fields, approaching at varying speeds, at times of considerable magnitude (of the order of 1 - 2.5 m/sec). It is therefore of interest to study the effect of the deformation rate of the value of ice's ultimate compressive strength, especially since the necessary clarity does not yet exist in regard to this question.

G.A. Pchelkin, in testing some samples of man-made ice, detected a decrease in the ultimate compressive strength at an increase in rate of stress increment (a, Kg/cm^2 per min), having developed the data presented in Table 8.

Table 8

Effect of Rate of Stress Increment on Ice Strength

Size of samples, cm	t, °C	a, Kg/cm^2 per min					
		10	12	16.5	17.5	21	36
7 X 7 X 7	-4	--	--	--	70	60	35
7 X 7 X 7	0	38	33	32	--	--	--

A similar pattern was obtained by Brown [164], also having significant decline in strength from 60 to 24 Kg/cm^2 for a high rate of stress increment (from 20 to 50 Kg/cm^2 per min).

N.A. Tsytovich [119] and S.S. Vyalov [35], studying the effect of the rate of load increase on the ultimate compressive strength of frozen soils, obtained a relationship of a contradictory type.

Let us point out that, in discussing the test results, N.A. Tsytovich writes: "With an increase in the rate of load feed, the temporary resistance increases, but at very high velocities, as experiments have shown in the study of tensile strength (at loads approaching dynamic ones in velocity), the temporary resistance again decreases somewhat".

Considering the effect of load rate on the ultimate compressive strength of ice, one should take into account certain features of the ice's mechanical properties, greatly complicating the pattern of the occurrence.

We therefore undertook experimental studies on the effect of deformation rate on the ultimate compressive strength of ice at deformation rates and temperatures close to those occurring under actual conditions of structural supports' functioning. Pressure was oriented perpendicularly to the axis of ice crystallization, in the manner that this took place in actuality.

Results of Experiments

From 1937 - 1950, the author undertook systematic studies of 477 ice samples of cubic form measuring from 7 X 7 X 7 to 20 X 20 X 20 cm. From 1960-1961, analogous studies were conducted by graduate student F.I. Ptukhin. The tests were run on a 30-ton Amsler machine at travel speed w of crossarm ranging from 2 to 25 cm/min.

A detailed description of the tests has been given in reports [21, 145]; therefore, we will list only the concise findings here.

1. At an increase in the rate of load feed, in all tests, the strength of the ice decreased considerably.

2. The degree of reduction in ice strength varied, as is apparent from Fig. 22, depending on temperature of the tests. As is known [126], the effect of the deformation rate on the resistance of materials to disruption will differ depending on test conditions and disruption mechanism.

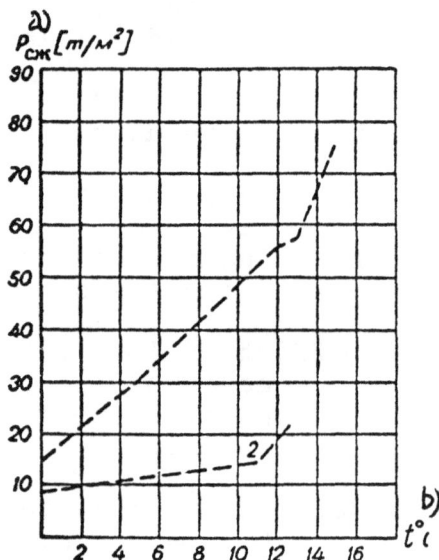

Fig. 22. Effect of Temperature and Deformation Rate on Ultimate Compressive Strength: 1 - at w = 2 cm/min; and 2 - at w - 20 cm/min. Key: a) $P_{compres}$ [tons/m^2]; b) t, °C.

At slow rates of deformation, the microdefects in the material's structure (cracks, hollows etc.) resulting from the shearing processes can be "healed" (closed), and thereby the material's strength will increase. At a faster rate of deformation, these processes do not manage to develop and the ultimate strength of ice to uniaxial compression decreases.

The breakdown of the samples at greater loading rates had a brittle nature; it occurred instantaneously, the sample crumbled into a pile of fine fragments, part of which jumped from the machine with force.

It is evident from Fig. 22 that the effect of the ice temperature was reflected less at high deformation rates. Such a dependence of the material's strength on temperature in case of brittle breakdown is generally known and ties in completely with the concepts of the nature of the materials' breakdown.

Substantiated conclusions can be made only on the basis of a fairly large number of observations when the effect of heterogeneity of individual samples in respect to their structure, dimensions, methods of processing etc. will not be reflected significantly on the magnitude of the average computed values.

For this purpose, for each cycle of observations, we computed $\bar{x}$ = arithmetic mean; σ = root-mean-square deviation; $m = \sigma /n$ = fundamental error of mean arithmetic value; $v = 100 \sigma/ \bar{x}$ = index of variation; and $\sigma_\delta = 0.707 m$ = fundamental error.

In addition, for a determination of the reliability of the difference in average values of each pair of cycles being compared (at w = 2 cm/min and w = 20 cm/min), we computed:

a) the difference of their mean arithmetic values:

$$\Delta x_1 - x_2 ;$$

b) fundamental error of difference in arithmetic means:

$$\sigma_{x_1 - x_2} = \sqrt{m_1^2 - m_2^2};$$

c) the ratio of fundamental difference $\Delta x_1 - x_2$ to the fundamental error of difference $\sigma_{x_1 - x_2}$.

It is known from the conclusions in the theory of mathematical statistics that if the difference between characteristics being compared is greater than 3 errors of the fundamental

difference, with a probability equalling 0.997, we can assert
that the difference obtained is actual and not fortuitous.

Table 9

Composite Table of Test Results

No. of cycle	w, cm/min	No.of samples	t, °C	x,Kg/cm^2	,Kg/cm^2	$\dfrac{\Delta x_1 - x_2}{\sigma_{x_1 - x_2}}$	probable relia-bility
1	2	15	-13.0	57.5	23.9	6.35	1.00
2	20	15	-12.8	16.2	7.9		
3	2	15	-13.6	55.1	7.8	7.41	1.00
4	20	15	-14.3	23.1	14.7		
5	2	11	-11.8	58.0	19.4	6.75	1.00
6	20	12	-11.0	16.9	5.8		
7	2	15	-16.0	60.0	11.3	14.4	1.00
8	20	15	-14.2	15.3	4.0		
9	2	13	-14.8	74.0	18.5	2.5	0.97
10	20	12	-13.5	48.5	30.8		
11	2	14	- 4.6	30.2	10.1	7.35	1.00
12	20	13	- 4.0	8.6	1.6		
13	2	14	0.0	14.8	4.0	3.31	0.99
14	20	12	0.0	9.8	3.6		
15	2	11	- 3.9	31.8	13.4	4.69	0.99
16	20	13	- 3.2	11.7	5.5		
17	2	13	-14.8	73.6	18.4	2.04	0.93
18	20	12	-13.5	54.3	27.4		
19	2	12	- 4.6	39.6	9.3	9.35	1.00
20	20	13	- 4.0	12.2	4.1		

Remarks. 1. Samples of cubical form, with edge length of
10 cm. 2. Loading was conducted perpendicularly to crystallization axis. Deformation rate fluctuated from 0.0033 to
0.0330 sec^{-1}.

Reviewing the data listed in Table 9, we become convinced
of the complete reliability of the results derived.

The allowance subsequently made for the dynamics of load
application showed that it is not simply by this factor alone
that we can explain such a significant decrease in ice strength.

A similar pattern was obtained by F. I. Ptukhin during tests from 1960-1961 on 213 ice samples at travel speed of machine's crossarm ranging from 2 to 32 cm/min (Fig. 23).

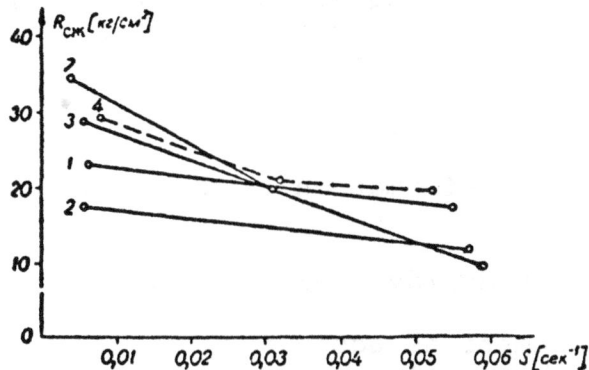

Fig. 23. Results of F.I. Ptukhin's Experiments.
Key: a) $R_{compres}$ [Kg/cm^2] and b) S [sec^{-1}].

Effect of Deformation Rate

Until now, we have spoken of the effect of the travel speed (w) of the machine's crossarm on ice strength at the given actual test conditions.

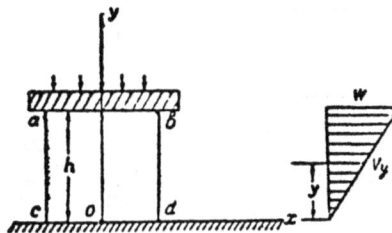

Fig. 24. On Determining the Deformation Rate
During Compression

It is more interesting to consider the effect of the deformation rate S which is determined not only by the travel speed of the machine's crossarm but also by the sample's dimensions, which permits us to use the test results in the transfer to natural conditions; rate of deformation can be computed according to the relationship:

$$S = dv/d\ell \ ,$$ (6.7)

where v = relative approach speed of 2 adjoining points of the sample; and ℓ = distance between them.

Taking into account that the velocity of the particles of material contiguous to the mobile crossarm over plane ab (Fig. 24) equals (w), while the velocity of the particles lying in the stationary plane cd equals zero, and assuming in a first approximation the linear law of distribution of rates of approach of the material's points, we find

$$v_y = w \ y/h \tag{6.8}$$

Obviously, the deformation rate in direction of the oy axis will equal

$$S = dv_y/dy = w/h \ sec^{-1}. \tag{6.9}$$

With the assumptions adopted by us, the deformation rate for all points of the sample proves to be a constant value, the magnitudes of which are given in Table 10.

Table 10

Dependence of Deformation Rate on Crossarm's Travel Speed

w, cm/min	2		10		20		25
h, mm	200	100	70	100	200	100	100
S, sec^{-1}	0.0017	0.0033	0.0048	0.0167		0.033	0.0417

As we shall observe, the deformation rate varied widely in our tests, i.e. by almost 25 times. The results of the investigation are illustrated in Fig. 25.

Unfortunately, in spite of the high total number of tests, for each temperature we obtained a total of 2-3 points, which is inadequate for explaining the nature of the relationship. Therefore the data of all tests are reduced to one temperature (-3° and 0°) [sic], which can easily be done by comparing the curves shown in Fig. 22.

These curves have been developed from extensive experimental material, carefully processed with the application of the methods of mathematical statistics and hence the reduction to one temperature can be done fairly reliably.

As a result of the calculation, we have obtained the values of ultimate strength of ice, referred to the temperature of -3° and 0°C, indicated in Table 11.

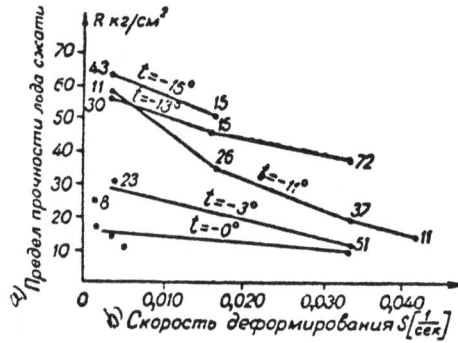

Fig. 25. Effect of Deformation Rate on Ultimate Compressive Strength of Ice. The curves reflect the number of tests and the ice temperature. Key: a) ultimate compressive strength of ice; b) deformation rate S [1/sec]; and c) R, Kg/cm^2.

Table 11

Effect of Deformation Rate on Value $R_{compres}$

w, cm/min	w, mm/sec	h,mm	S,sec^{-1}	$R_{compres}$,Kg/cm^2		No. of samples	Remarks
				t = -3°C	t = 0°		
2	0.33	200	0.0017	26,0	15,6	22	Ob' River
		100	0.0033	23.4	13.9	119	Ice, loaded
		70	0.0048	17.2	11.3	22	perpendicular-
							ly to crystals'
							axis
10	1.66	100	0.0167	11.8	8.4	70	Temperature re-
20	3.33	200					duced to 0° and
		100	0.0333	9.9	8.2	130	3°
25	4.17	100	0.0417	8.8	7.5	11	

Analyzing the data obtained, we conclude:

1. The deformation rate is reflected significantly on the ultimate compressive strength of ice. The increase in the rate reduces the ice strength (in the range of S investigated).

2. The law of variation in ultimate strength at temperature of -3°C and deformation rate of the order of 10^{-3} - 10^{-2} sec is described satisfactorily by the relationship:

$$R_{compres} = \frac{3 \cdot 10}{\sqrt[3]{S}} \quad Kg/cm^2 , \qquad (6.10)$$

and at a temperature of 0°, by the expression

$$R_{compres} = \frac{2.50}{\sqrt[3]{S}} \quad Kg/cm^2 \quad . \tag{6.11}$$

The relationships derived can be utilized for an explanation of the ice's strength in the period of spring ice passage. For this purpose, it is necessary to know the floe's rate of deformation in the zone of interest to us.

It is worth noting that under actual conditions, the deformation rate in various points of the zone contiguous to the contact surfaces of the supports and the floe (here the deformations will be plastic) will vary. Obviously the deformation rate will be maximum at the face of the ice apron and minimum at the boundary with the elastic zone.

As a first approximation, it is possible to be limited to a consideration of the average deformation rate in the plastic zone.

Based on our tests on the introduction of punches into ice, we can adopt an area of plastic deformations in a floe extending for the distance H = 1.5 - 2 widths of support b_0. At floe's travel velocity v in m/sec and propagation depth of plastic deformations H = $2b_0$, we get the average deformation rate in the limits of the plastic zone

$$S = w/H = v/2b_0 \, sec^{-1}. \tag{6.12}$$

The S-values computed in such a way for various conditions are listed in Table 12.

Table 12

Value of Deformation Rate

Rate of floe movement, m/sec	Width of support, m		
	3	4	5
	S, sec^{-1}		
0.25	0.042	0.031	0.025
0.50	0.083	0.062	0.050
1.00	0.166	0.125	0.100
1.50	0.25	0.187	0.150

It is evident from the table that the deformation rates occurring in the period of displacements (v_o = 0.25 m/sec, S = 0.060 sec^{-1}) have been encompassed by our studies. However, the deformation rates during motion of large ice floes (v = 0.5 - 1.0 m/sec) although they go beyond the limits of those observed, do connect directly to them, in connection with which we consider it possible to extrapolate Eqs. (6.11) and (6.12) obtained by us, to the area of operation of the supports' ice aprons (starlings). We thus have:

$$R_{compres} = \frac{2.50}{\sqrt[3]{S}} = 3.15 \sqrt[3]{b_o/v} \text{ Kg/cm}^2 .$$

Consequently:

1. The deformation rate has an appreciable effect on the ultimate compressive strength of river ice. With an increase in the deformation rate, the ultimate strength decreases (at S-value greater than 0.01 sec^{-1}).

2. Based on the extensive experimental material obtained by us, we are able to establish the following relationships between deformation **rate S and** ice's ultimate strength $R_{compres}$:

$$R_{cx} = \frac{3,10}{\sqrt[3]{S}} \ (t = -3° C);$$

$$R_{cx} = \frac{2,50}{\sqrt[3]{S}} \ (t = 0°.C)$$

at S-values ranging from 0.0017 to 0.00417 sec^{-1}.

3. The deformation rates of an ice field occurring during the functioning of the structures' supports fall within the limits of 0.025 to 0.250 sec^{-1}, i.e. they directly adjoin the ranges studied by us and even partly overlap them.

In connection with this, it is admissible (in the first approximation) to have a certain extrapolation of Eqs. (6.10) and (6.11) in considering the conditions of the supports' operation.

4. The effect of the velocity of motion v of the floes on ultimate compressive ice strength can therefore be estimated with the relationship:

$$R_{compres} = 3.15 \sqrt[3]{b_o/v} \text{ Kg/cm}^2$$

which at the frequently occurring values of supports' width b_0 amounting to 3-5 m can be replaced by the simpler expression:

$$R_{compres} = 5/\sqrt[3]{v} \quad Kg/cm^2 \quad (v > 0.01 \ m/sec). \qquad (6.13)$$

5. In the presence of weak ice having been subjected to intensive spring thawing, the strength can be reduced by about twofold and is computed with the relationship:

$$R_{compres} = \frac{2.50}{\sqrt[3]{v}} \quad Kg/cm^2 \qquad (6.14)$$

(at b_0 = 3 - 5 m).

5. Ultimate Strength of Spring Ice During Flexure

Strength of spring ice during flexure has been studied by many researchers, chiefly Russian and Soviet. Of special interest are the observations of the strength of a natural ice cover. Let us cite certain data:

1. I.P. Troshchinskiy, having tested under natural conditions 40 "keys" from the ice sheet on Lug R., obtained the following values for the ultimate bending strength of ice during flexure by a force directed:

upward.......... R_M = 7.1 Kg/cm^2

downward........ R_M = 14.0 Kg/cm^2.

2. A considerable decline in ice strength during bending up to the period of spring ice-out was recorded by Yu.N.Neronov [28], having obtained for "keys" from the greatly weakened ice on the Neva R. on 14 March 1943 a value for ultimate bending strength of ice equalling a total of only

$$R_M = 4 \pm 1 \ Kg/cm^2,$$

which is about half as much as the values normally found.

3. Analogous results were obtained by N.D.Shishov [127], having conducted tests of 45 ice samples on the Northern Dvina R. at Arkhangel'sk from 1938-1943 and having obtained an ultimate bending strength of 2.5 Kg/cm^2.

N.D. Shishov concludes that the ice strength under natural conditions is several times less than the strength derived during the testing of ice blocks.

I.P. Butyagin [30] conducted many observations of the breakdown of ice supports and beams of large dimensions (of the order of 2 X 0.8 m in plane) cut in the natural ice sheet of a river and a reservoir. He notes that the ultimate bending strength of ice in the pre-ice passage period does not exceed 5.50 Kg/cm^2; the strength of the ice sheet during upward bending is slightly less than during downward bending.

The value of ultimate bending strength of ice is usually established as the maximal tensile stress in the ice sample, being bent, prior to its breakdown, computed on the basis of the formulas for materials' resistance, given for an elastic body.

Since the bending of ice is also always accompanied by plastic deformations, the data developed obviously are approximate and probably slightly exaggerated.

Results obtained by various authors for temperatures close to zero are given in Table 13.

From an analysis of Table 13, it follows that:

1. The ultimate bending strength of river ice under laboratory conditions, at temperature of 0° and static load, will fluctuate within fairly narrow limits of 9 - 12 Kg/cm^2.

2. In the tests conducted on ice under natural conditions by the method of "keys" being bent by an upward directed force, the strength limit is lowered to 5 - 7 Kg/cm^2. The decrease in the R_H -value is explained mainly by the larger number of fissures and inclusions in the large ice masses.

3. As was demonstrated by Yu. N. Neronov, I.P. Butyagin and A.R. Shul'man, at a high degree of breakdown of the ice cover, the ultimate strength falls to 3 - 5 Kg/cm^2, i.e. by about 1.5 - 2 times.

Let us consider the effect of deformation rate on the value for ultimate bending strength of ice.

For a clarification of this question, we tested 40 blocks with a size of 10 X 10 X 30 cm, some of which were broken down at travel speed of machine's crossarm amounting to 2 cm/min (25 blocks), while the remainder were broken down at travel speed

Table 13
Ultimate Bending Strength of Spring Ice

Researcher	Year	t, °C	No. of samples	Rate of loading	Type of ice	R_H Kg/cm^2	Source
B.P. Veynberg...	1911	0	12	static	river	11.1	[12]
B.P. Veynberg...	1913	0	25	"	"	12.8	[12]
B.N. Sergeyev...	1921	0	15	"	"	11.5	[15]
V.N. Pinegin....	1933	-3to-5		"	"	18.0	[14]
Yu.A. Pedder ...	1925	0		"	"	8.5	[106]
M.M. Basin......	1934	0	8	"	"	11.8	[16]
K.N. Korzhavin..	1937	0	25	2 cm/min	"	9.2	[19]
K.M. Korzhavin...	1937	0	15	20 "	"	3.6	[19]
I.P. Troshchinskiy	1939	0	40	static	"	7.1	[12]
F.F. Orlov......	1940	0	--	"	"	14.0	[168]
N.D. Shishov....	1942	-2	6	"	"	10.5	[170]
N.D. Shishov...	1940-42	-1	4	"	"	2.5	[170]
Yu.N. Neronov...	1943	0	10	"	"	4+1	[28]
P.A. Pisarev....	1944	0	10	"	"	9.8	[169]
B.D. Kartashkin..	1945	-2	1	"	"	12.8	[24]
A.R. Shul'man...	1945	0	2	"	lake	4.0	[25]
P.V. Fedorov....	1941	-1	8	"	river	14.7	
I.P. Butyagin...	1953-59	0	31	"	"	3.7	[30]
I.P. Butyagin...	1958-59	0	25	"	reservoir	6.2	[30]

of 20 cm/min. The tests were run on a general-purpose Amsler machine at temperature of 0°C. The test data indicated that in the case of bending at an increase in loading rate, the ice strength is reduced by about 2 - 2.5 times.

In a first approximation, we can consider the following relationship valid:

$$\frac{R_{20}}{R_2} = \sqrt[3]{\frac{w_2}{w_{10}}}.$$

(6.15)

In addition, based on Yu. N. Neronov's data [28], we constructed a graph of the dependence of ultimate bending strength of ice on the rate of increase in stresses, indicating that an increase in the loading rate leads to a decrease in the ice's strength.

6. Ultimate Strength of Ice Under Shearing Stress

The question concerning the ultimate shearing strength of ice has been studied less than the others, partly in connection with the difficulties involved in the procedure for conducting the tests. We are cognizant of the carefully formulated experiments by V.N. Pinegin [14], M.M. Basin [16], and I.P. Butyagin [32], having conducted 73 observations of the cutting of an ice beam formed in the depth of the ice pack on a Siberian river. Let us note that I.P. Butyagin's studies were conducted at the end of March - beginning of April over ice, the structure of which (as a result of the spring thawing processes) still did not vary appreciably.

Table 14

Ultimate Shearing Strength of Ice

Researcher	Year	t, °C	Direc. of force	Ultimate strength	type of ice	Remarks
V.N. Pinegin	1922	0-2	parall.	6.5	river	
Finlayson	1927	1.1	"	6.9	"	
M.D.Sheykov &	1930	0	"	9.9	artif.	
N.A.Tsytovich		-0.4		11.0	"	
		-2.9		27.4	"	
M.M. Basin	1934	0				
		0		2.82	"	aver.of 102 samples
		0		1.50	"	aver.in period of ice passage
B.D.Kartashkin	1947	-10.5		6.8	poured in	based on torsional tests (3 samples)
B.N. Veynberg	1940	--		5.8	--	aver. of 111 tests
P.V. Fedorov	1941	from -1° to	"	6.0	river	16 tests
		-2°	perpend.	5.3	"	11 tests
I.P. Butyagin	1956-57	0	parall.	4.5	"	
I.P. Butyagin	1956-57	0	"	2.2	"	before ice-out

Therefore there is no doubt that under the conditions of structural piers' functioning during the ice-out period, the ultimate shearing strength of ice can be less (Table 14).

9

As we shall observe, the shearing resistance of river ice will fluctuate (at temperatures from 0 to -2°) within fairly narrow limits of 5-7 Kg/cm^2.

It is important to note that according to studies conducted by M.M. Basin [16], the ultimate shearing strength of ice on the Svir' R. by the time of ice passage in 1934 dropped considerably and could be computed with the formula:

$$\tau = \tau_0 - \left(\frac{\Sigma t}{29}\right)^2, \qquad (6.16)$$

where τ_0 = ultimate shearing strength of ice prior to onset of steady warming; and Σt = sum of mean diurnal air temperatures from time of steady warming.

M.M. Basin indicates that in the period of ice passage, the ultimate strength fell from 2.8 Kg/cm^2 to 1.5 Kg/cm^2, while in isolated cases it fell to 0.75 Kg/cm^2.

As a first approximation, we can thus consider that in the period of spring ice passage, the ultimate shearing strength of ice can be of the order of:

a) for the rivers of the North and Siberia, in which the ice strength up to the time of opening does not manage to decrease greatly,

$$R_{av} = 5 - 6 \text{ Kg/cm}^2;$$

b) for the rivers of the European sector of the USSR, in which the ice strength by the time of ice passage has greatly decreased,

$$R_{av} = 1.5 - 3 \text{ Kg/cm}^2.$$

It is worth noting that in connection with the low degree of study of the question, the undertaking of additional investigations to establish the ultimate shearing strength of the ice has great significance in solving many practical problems.

7. Ultimate Strength of Spring Ice During Tensile Strain

The ultimate tensile strength of spring ice has been studied relatively little. The data at our disposal concerning a determination of the ultimate tensile strength of ice at temperatures close to zero are reflected in Table 15.

Table 15

Ultimate Tensile Strength Of Ice

Researcher	Year	t, °C	Direc. of force	Ultimate strength Kg/cm^2	Nature of Ice	Number of samples
B.P. Vasenko	1897	-4		12.3	artif.	3
V.N. Pinegin	1922	0 to -2	perpend.	7.5	river	200
				10.0	"	
B.P. Veynberg	1940	-3		11.1	"	235
B.D. Kartashkin	1947	-3	"	10.3	"	
		-3.5		11.4	poured	
Kirkgam	1927	0 to 5		6.7	river	

As we shall observe, the closest to the natural conditions of the piers' operation are the tests by V.N. Pinegin, in accordance with which

$$R_p = 7 - 10 \text{ Kg/cm}^2,$$

There are bases for assuming that for weak ice, the ultimate strength (perpendicular to the crystals' axis) does not constitute more than $3 - 4$ Kg/cm^2.

8. Effect of Sample's Dimensions

The question of the effect of a sample's dimensions on the material's ultimate strength has been a topic of research for many years.

O.M. Gumenskaya [129], N.A. Tsytovich and M.I. Sumgin [119], having studied the ultimate compressive strength of frozen soils, noted that with an increase in the samples' dimensions, the ultimate strength increases.

A similar nature of the dependence (for cubes with an edge of 10 and 20 cm) was noted by the author [19] for river ice. Tests made on 161 specimens of selected ice indicated that the ultimate compressive strength of river ice also increased with an increase in the sample's dimensions, and only in one instance was the dependence reversed; this was explained by the less uniform structure in the large ice samples.

It is thus necessary to reckon with two factors influencing the ultimate strength of ice. With an increase in a sample's dimensions, the ice's ultimate strength (just as of many other materials) increases and can simultaneously decrease in connection with the greater heterogeneity in the structure of the large ice samples.

In V.V. Lavrov's report [36], he presents data on a decrease in the ultimate bending strength of ice with an increase in its thickness, and he expresses the opinion concerning the nature of the "scale effect", having served (at the initiative of N.N. Davidenkov) as a topic of discussion of the pages of the journal Zavodskaya Laboratoriya (Plant Laboratory).

The extensive research undertaken from 1953-1961 in the Laboratory of Ice Thermics at the Siberian Branch of the USSR Academy of Sciences under I.P. Butyagin's supervision permitted him to demonstrate convincingly in a series of articles [30, 128, 130] that under natural conditions of the occurrence of an ice pack, its ultimate strength declines with an increase in the dimensions of the sample.

The most probable cause for this consists in the greater heterogeneity in the structure of the large ice samples and the related concentration of local stresses during the tests.

Table 16

Ultimate Bending Strength of Ice, Kg/cm^2 [See Note]

Nature of ice	Year	Dimension of sample's cross section, cm				No. of Identifications
		8X8	55X70	80X50	80X70	
River.........	1953-1957	12.0	--	--	3.7	73
River.........	1953-1957	10.3	2.9	--	--	42
Reservoir.....	1958-1959	11.0	--	6.2	--	66

([Note]: For forces parallel to the crystal's axis and at temperatures close to 0°C.).

The quantitative aspect of the question is illustrated by the following collated data (Table 16).

I.P. Butyagin's investigations showed that the extent of varying the ultimate strength of an ice sheet decreases with an increase in the dimensions of the section, as is clear from Fig. 26. It is interesting to consider the data, discussed by I.P. Butyagin and the author [31] concerning a determination of the deformations and strength of 4 ice fields, brought to the point of breakdown by vertical concentrated forces.

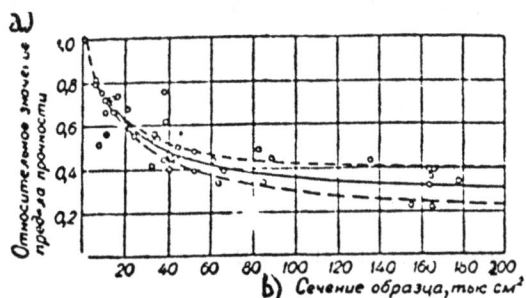

Fig. 26. Effect of Samples Sectional Area on Ultimate Bending Strength of River Ice (after I.P. Butyagin's tests). Key: a) Relative value of ultimate strength; and b) Section of sample, thous.cm^2.

For the purpose of an experimental verification of the theoretical assumptions and also for establishing the actual values of the ultimate strength of an ice sheet without the preparation of samples, on the Novosibirsk Reservoir, we tested four ice fields which were being loaded at the edge by a vertical concentrated force. For conducting the test on the ice sheet of the reservoir, we selected a smooth sector remote from the shores, in which we sawed a narrow straight cut with a length of 100 m.

In its central part, on the edge of the ice, we mounted two round tanks each with a capacity of 5 m^3. The load was developed by filling the tanks with water, according to the volume of which we established the amount of force.

The saggings in the ice sheet were recorded simultaneously at all points after each load increase of 500 kg. In this way, each experiment permitted us to obtain a series of curves for the deformation of the ice field, corresponding to the calculation system adopted. As could have been expected, the maximum deformations occurred at the point of applying the load (in the middle of the cut) and attenuated relatively quickly away from it. Their maximum values prior to breakdown comprised from 80 to 105 mm. The disruption of the ice fields at finite values of the load developed suddenly, as usually occurs in the tests for the strength of ice specimens.

The fracture of the ice pack occurred along one curvilinear fissure as a semicircle, having bent around the point of load application and having separated from the ice field a chunk of fairly regular form. In this way, there was no ice fracture at the points of applying the loads and the loading tanks stayed on the broken-off chunk, without sinking. Simultaneously with the tests which were conducted on the ice fields, in these same sectors of the ice sheet, we tested the beam console samples ("keys") for bending.

We have presented in Table 17 the basic data from these tests.

Table 17

Results of Tests Conducted on Ice Fields and Console Samples of Ice Pack

No.of test	Date of test	t,°C	Average ice thickness, cm	Max. load on edge of ice field Kg	Ultimate strength, Kg/cm^2	
					of field	of console
1	20 Dec.1957	−7.0	44.0	72.20	3.4	7.5
2*	4 Dec.1958	−3.0	35.0	5500	--	--
3	9 Dec.1958	−2.8	32.0	5880	5.0	7.6
4	17 Dec.1958	−8.5	40.0	8490	5.4	8.0

*Disruption of ice field did not occur.

In summing up, we can conclude:

1) with an increase in dimensions of ice samples, a tendency develops toward an increase in the ultimate strength (just as of certain other materials);

2) there is a simultaneous increase in the degree of hetero geneity of the ice pack, leading to a concentration of stresses and a premature breakdown;

3) with an increase in a sample's dimensions, the degree of decline in ultimate strength decreases, tending toward a certain limit; and

4) full-scale investigations permit us to establish that the ultimate bending strengths of ice blocks, consoles and ice fields are about as follows:

block (8 X 8 cm) --1.00; console (80 X 50 cm and 60 X 70 cm) --0.33--0.50 respectively; and ice field --0.25-0.35.

9. Recommended Values of Ultimate Strength

In a determination of the ice pressure on structures, it is natural to proceed from its maximal values. In this connection, however, one should take into account a logical reserve of the structure's strength and not consider the cases of a catastrophic nature.

Generalizing the information discussed above, let us present a summary of the essential determinations for the ultimate compressive, crushing, bending, tensile and shearing strength of ice. In this connection, we will have in mind that in respect to the strength of an ice field during the ice breakup period, most of the USSR rivers can be classified under one of the following groups:

a) rivers opening chiefly owing to the breakup of ice, still fairly stable, by flood waters (large rivers in Siberia, flowing from south to north, and certain rivers in the North); and

b) rivers, the opening of which occurs after an intensive thawing of the ice pack, having managed to lose a considerably amount of its strength (most of the rivers in the European sector of the USSR, certain rivers in the south of the Asiatic sector of the USSR).

In connection with this, we should adopt various values for the calculated strength of ice from each group of rivers. The composite data on ice strength are listed in Table 18.

Table 18
Ultimate Strength of Ice, t/m^2

Nature of deformation	Notation	Rivers	
		North & Siberia	European part of USSR
Compression....	$R_{compres}$	100 - 120	50 - 70
Bending	R_M	90 - 120	50 - 70
Local crushing.	R_{cM}	250 - 300	125 - 175
Shearing	R_{cp}	50 - 60	15 - 30
Tensile strength	R_p	70 - 100	30 - 40

As we shall observe, the ultimate compressive and bending strengths are about the same.

Examining the question of the effect of deformation rate, we established that the ultimate compressive and bending strength depend considerably on deformation rate S. The results of the tests analyzed above were usually obtained at loading rate w = 2 - 5 cm/min which at dimensions of specimens being compressed amounting to 7 - 10 cm corresponds to a deformation rate ranging from 0.0033 to 0.0083 sec^{-1}. Therefore the ultimate strength limits presented in Table 18 should be referred to a deformation rate about equal to 0.0060 sec^{-1}, while for the other rates, one should conduct a conversion according to the dependence:

$$R_{compres} = \frac{2.50}{\sqrt[3]{S}} \quad Kg/cm^2 \quad (6.11)$$

The technique for the calculations of ultimate strength ofice during compression (width of support b_0 = 4 m) is given in Table 19.

Table 19

Ultimate Compressive Strength of Ice

S, sec^{-1} (after 6.9)	v, m/sec (after 6.12)	$R_{compres.}$ t/m^2 after 6.11)
0.006	0.05	137
0.062	0.50	62.5
0.125	1.00	50
0.187	1.50	45

After a certain rounding off, it is possible to suggest the following values of calculated ultimate strength (Table 20).

These values are recommended for a fixed orientation of force relative to the crystal's axis and for ice temperatures close to 0°C.

Table 20

Recommended Values of Ultimate Strength of Ice, tons/m^2

Nature of deformation	Direction of force	Notation	Rivers of North & Siberia			Rivers of European part of USSR		
			mobile, v=0.5 m/sec	full-scale ice out		mobile v=0.5 m/sec	full-scale ice-out	
				v=1 m/sec	v=1.5 m/sec		v=0.5 m/sec	v=1.5 m/sec
Compression	perpend.	R_{comp}	65	50	45	40	30	25
Local crushing..	"	R_{cM}	150	125	110	80	65	55
Bending..........		R_M	65	50	45	40	30	25
Shearing........	parall.	R_c	--	40-60	--	--	20-30	--
Tensile strength.	"	R_p	--	70-90	--	--	30-40	--

Table 21

Coefficients for Transfer to Compressive Strength of Ice at
t = 3°C

Ice temperature, °C						Remark
0	-3	-6	-10	-12	-15	
1.15	1.00	0.92	0.83	0.80	0.78	After Veynberg
1.68	1.00	0.74	0.54	0.46	--	After Korzhavin for w = 2 cm/min
1.06	1.00	0.84	0.63	0.50	--	After Korzhavin for w = 20 cm/min

At lower temperatures of ice, it is necessary to make a conversion either based on the relationship:

$$R_1 = 11 + 3.5 \, t, \; Kg/cm^2 \qquad\qquad (6.1)$$

or based on the conversion coefficients of strength (Table 21) at some given temperature suggested by B.P. Veynberg [1] or by the author [19].

Chapter VII

EFFECT OF ABUTMENT'S FORM IN PLANE

1. Basic Concepts

Importance of Question. In the planning of bridges and hydraulic engineering structures on rivers with severe ice passage conditions, the piers are usually given a tapered form, facilitating the breakdown of the ice sheet and developing smooth flow around the supports.

Undoubtedly the variation of a pier's form in plane (or a variation in the taper angle of the ice-cutting face) can lead to a significant variation in the nature of stressed state and hence in the conditions of the ice pack's breakdown.

Operating Features of Piers with Various Forms. Let us indicate with some examples the operating features of supports having varying form.

a) At introduction of a punch of rectangular shape into the ice (or into some other linearly-stressed medium), pressure distribution on the contact plane will be very uneven at the initial moment.

As is known from the theory of elasticity, the maximum compressive stresses develop at the punch's edges, where in connection with this, areas of plastic flow appear, or the material's brittle breakdown develops. At an increase in pressure P, the dimensions of the areas of plastic deformations increase and finally merge; they may even overlap one another, thus forming a zone of increased pressure.

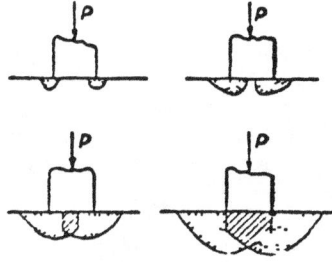

Fig. 27. Formation of Compactions of Zone at Intrusion of Punches.

The compacted zone (cross-hatched in Fig. 27) constitutes
as it were a natural extension of the punch. The material within
it exists under conditions of high multilateral compression and
in the shearing planes, considerable compressive forces are
acting. As tests show [131], under high compression conditions,
the shearing resistance increases by tens and even hundreds of
times, so that the flow of material in the compacted zone becomes difficult. The subsequent effect on the material is actually not achieved directly but through the compacted zone, the
form of which thereby acquires great importance.

We have shown in Fig. 28 the form of the compacted zone
having formed in the ice sample at injection of the rectangular
punch. The latter, acting through the compacted zone, proved
to be equivalent to a triangular punch and broke the specimen
into two parts.

b) At introduction of punches of triangular shape, the
formation of a compacted zone also occurred as is obvious e.g.
from Fig. 29, illustrating one of the experiments. However, for
punches of triangular shape, the form and dimensions of the
compacted zones will vary with a variation in the angle of
tapering 2α.

As follows from Ratier's [132] quite interesting experiments on the cutting of sand, at the sharper piers, the dimensions of the compacted zone become less and less and finally at
$2\alpha < 60°$, the zone is no longer formed. Let us point out that
at a reduction in the angle of tapering, the form of the compacted zone more closely resembles the punch's shape.

We can thus consider that the tapering angle of the pier
has a significant effect on the form of the compacted zone and
consequently on the amount of the material's resistance to
deformation also.

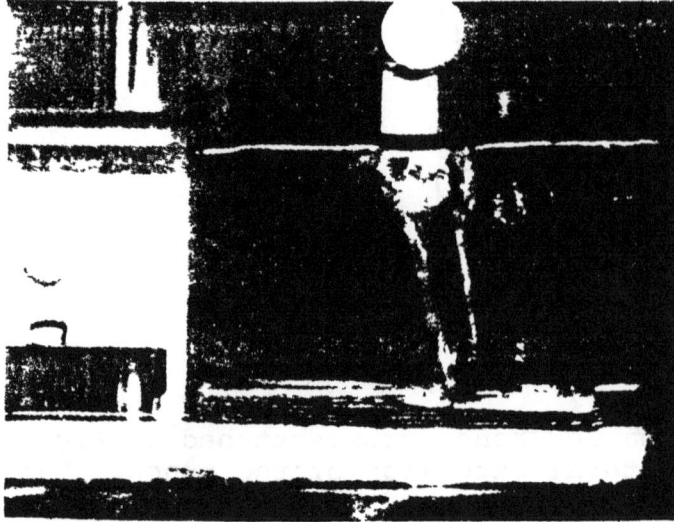

Fig. 28. Nature of Phenomenon at Pressing of Rectangular Punch Into Ice Sample.

Fig. 29. Nature of Phenomenon at Imbedding of Triangular Punch Into Ice Specimen.

The reason for the variation in dimensions of the compacted nucleus is found in the variation in the nature of the stressed state which specifically confirms the arrangement of isochromes obtained by A.N. Zelenin [132] in a gelatin plate, occurring in polarized light with a rectangular punch and a triangular one with a point angle of 60°. Such an occurrence can also be noted in the cutting of soils.

We can thus consider it established that the variation in the punch's shape in plane can vary significantly the nature of the material's stressed state and by the same token can be reflected on the value of the originating forces of interaction; the punches with a "blunt" shape cause a local crushing of the material, and the "sharper" ones cause its splitting; the total force required for the introduction of the punch includes surmounting the drag (less for the sharper punches) and frictional forces on the lateral faces (greater for the sharper punches).

Therefore an optimal punch form exists, requiring the least forces for its introduction into the material, which corresponds completely to the findings of the plasticity theory.

Considering the process of intrusion of a rigid convex punch into a plastic medium and assuming that friction between the punch and the medium is lacking, we have [111] the value of the required pressure in the form:

$$P = 2ka (\pi + 2 - 2\gamma)h, \qquad (7.1)$$

where h = thickness of material's layer; k = plastic constant;

$$k = \frac{\sigma_s}{\sqrt{3}} = \frac{2+\sqrt{3}}{2\sqrt{3}} \tau_{max}$$

σ_s, τ_{max} = flow point and maximum tangential stress; α, γ = values of angles shown in Fig. 30.

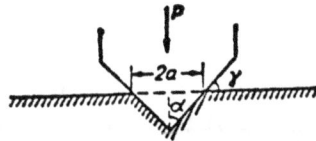

Fig. 30. On Establishing the Effect of Rigid Punch Entering a Plastic Material.

Our experiments on the crushing of 54 samples of river ice provide the possibility of roughly determining the plastic constant for river ice;

$$k = \frac{2+\sqrt{3}}{2\sqrt{3}} \frac{R_{cж}}{2} = 0,54 R_{cж}, \qquad (7.2)$$

where $R_{cж}$ = ultimate compressive strength of ice.

Considering the moment of the punch's insertion for its entire width b_o, we find

$$a = 0.5 \, b_o; \quad \gamma = 90 - \alpha,$$

after which Eq. (7.1) can be given the following form:

$$P = nR_c \text{✳} \, b_o h, \tag{7.3}$$

where $n = 1.08 \, (1 + \alpha)$, $\tag{7.4}$

2α = punch's tapering angle expressed in radians.

Results obtained from calculating the n-values for various tapering angles 2α are listed in table 22.

Table 22

n and m Values

2α	180°	120°	90°	75°	60°	Remarks
n	2,73	2,21	1,93	1,79	1,65	(after 7.4)
n	100	81	71	66	61	in %
m	1,00	0,81	0,71	0,66	0,61	(after 7.5)

In this & other tables, commas are
equivalent to decimal points.

Assuming that an increase by 2.73 times in the ultimate compressive strength of ice at introduction of a square punch is explained by the local crushing effect, let us find the coefficient taking the punch's form into account,

$$m = \frac{n}{2.73} \, 0.39 \, (1 + \alpha), \tag{7.5}$$

the values of which are also listed in the table.

The previous conclusions are partly based on investigations of the process of cutting soils and on certain assumptions in the theory of plasticity.

As is known, ice has a unique structure and a significant anistropy (reflected quite markedly at temperatures close to zero) and also is often subject to brittle breakdown, in connection with which the ice breakdown process can differ somewhat from that described above.

It thus proved useful to undertake a study of the effect of the pier's form in plane on the resistance of the ice sheet.

2. Studies by the Author

Test Conditions

For an explanation of the effect of the pier's form in plane on the value of the interaction forces (originating under actual conditions) between the pier and the floe, we determined the ice's resistance to insertion of punches having a varying form.

In order that the test conditions would be as close as possible to the natural ones: a) we used natural ice from the Ob' R.; b) we used samples of the largest sizes (based on dimensions of the testing machine) in the form of plates measuring 200.120.70 mm; c) the loading was conducted perpendicularly to the crystal's axis; and d) samples' temperature close to 0°C was maintained.

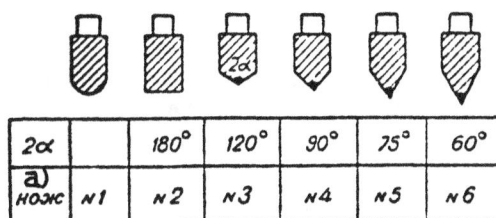

2α		180°	120°	90°	75°	60°
a) нож	N1	N2	N3	N4	N5	N6

Fig. 31. Form of Punches Utilized During the Investigations. Key: a) edge(face).

The tests were conducted in the laboratory of the Novosibirsk Institute of Railway Transport Engineers on a machine making it possible to obtain a compression diagram on a very large scale. The travel rate of the machine's crossarm was constant and equalled 1.5 cm/min. For the testing, use was made of 6 punches (blades), the dimensions and form of which can be judged from Fig. 31.

Let us note that, desiring to develop conditions similar to the cutting of supports into a floe of large dimensions (in which the deformations of material to one side become difficult), we tested many specimens under lateral compression conditions. In this connection, the displacement of the sample's ends to one side became almost impossible. In all, 73 samples were tested (Table 23).

Table 23

Description of Tests

Test conditions	Type of Blade						Total
	1	2	3	4	5	6	
Without lateral compression......	4	5	2	2	6	2	21
With lateral compression......	5	22	5	5	10	5	52
Total.....	9	27	7	7	16	7	73

Nature of Samples' Breakdown

The nature of the samples' breakdown naturally vary depending on local conditions and primarily on the type of blade. Let us consider the breakdown conditions in each separate instance.

a) Blades of semicircular shape. In the initial period of introduction of blade, plastic deformations occurred, resulting in the appearance of a sagging on the sample's surface, corresponding to the blade's form. Then, at further increase in pressure, on the transparent ice sample, rounded outlines of the compacted zone became visible.

If the sample had not experienced lateral compression, its splitting would have quickly developed, in which the direction of cracks always deflected by 9 - 10° from the blade axis to one side or the other.

However, in the event that the lateral compression had obstructed the sample's deformation, the breakdown would have occurred as follows. In the central part of the sample (under the compacted zone), small web-like fissures would have appeared (traces of intercrystalline displacements), the number of which would then have increased--the dimensions of the compacted zone would have grown. Then the ice block cracked, which in individual cases was accompanied by the emergence of the slip planes to the upper edge of the sample. In this connection, the cracks always diverged by a wider angle than in the case of free breakdown (without lateral compression).

b) Blades of rectangular shape. With blades of rectangular shape, the pattern of breakdown was similar to that just described

The compacted zone had the rounded form depicted in Fig.28. In the absence of lateral compression, the sample quickly broke up from fracturing, wherein the fissure deviated from the blade's axis by 5 - 14°. Under the conditions of constrained deformations (of lateral compression), the compacted zone gradually expanded and moreover many cracks appeared; some of them deviated from the sample's axis by 45-60° and even emerged to the end surfaces of the sample.

c) Blades of triangular section. The introduction of triangular blades is also accompanied by the appearance of a compacted zone of rounded form, clearly distinguishable in Fig.29. In the absence of lateral compression, the specimens rapidly cracked up, wherein the fissures' direction varied with a variation in the tapering angle, from 2° to 14°.

At lateral compression (after the attainment of a compacted zone of certain dimensions), the successive formation of two and even three fissures took place.

3. Discussion of Results

In the testing process, for each sample, we obtained a diagram reflecting the ice's resistance to deformation and the depth of the blade's penetration. The testing machine's features permitted us to obtain a large-scale diagram, owing to which it became possible to determine the resistance to deformation at any moment, with sufficient accuracy. The maximum forces required for breakdown of ice at introduction of punches at various forms into it are indicated in Table 24.

As we see from Table 24, the effect of the sample's form on the ice's resistance proved quite appreciable, which by the way was to have been anticipated.

The appearance of the compacted zone of circular shape for a square punch and features of triangular punches' form leads to the development of forces H according to the pattern in Fig. 32. It is obvious that the narrower the tapering angle of the support (2 α), the greater the horizontal component H of force and hence the more likely the splitting of the floes, often having been observed by us.

Table 24

Force in Kg and % Required to Crush a Sample

Test conditions	Type of Blade						Remarks
	rectang. $2\alpha = 280°$	semicir- cular	triangular with angle 2α				
			120°	90°	75°	60°	
With lateral compression	1093	943	900	705	588	445	Kg
	116	100	95	74	62	47	%
Without lateral	343	330	320	270	--	158	Kg
compression	104	100	97	82	--	48	%
Number of samples tested	11	25	9	9	10	9	total:73

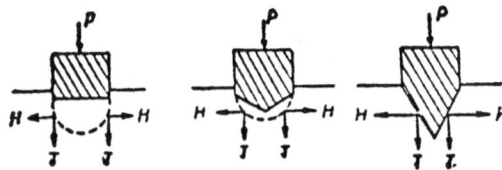

Fig. 32. Nature of Interaction of Punches Having
Varying Shape, with Varying Material.

Naturally the effect of lateral compression altered the
form of stressed state and, at the same pressure, led to a decrease
in the value of tangential stresses (responsible for the breakdown) and hence to the requirement for a further increase in pressure P on the punch.

The nature of variation in cutting forces becomes clear
from the diagrams obtained for all 73 specimens tested. In Fig.33,
we have shown the curves for the cutting forces obtained in testing
samples for all the blade types reviewed by us. In the figure,
we have indicated the times of the first crack's appearance.

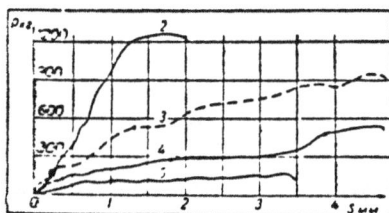

Fig. 33. Typical Stress Curves for Ice During Insertion of Punches. Numbers refer to the type of
punch, while the dot shows the time of fissures'
appearance.

Initially, there occurs the formation of a compacted zone,
wherein the resistance to penetration increases rapidly, about
in accordance with the linear law (segment OA, diagram in Fig.
11). We then note a more gradual increase in the cutting forces,
linked with the propagation of the region of residual stresses
into remote zones of the material (segment A Б). Finally, cracks
appear; after their formation, the deformations increase at
virtually unchanged resistance of the material. The cracks
spread gradually into the depth of the ice, imparting a stepped
appearance to the curve and finally, the crack reaches the
sample's edge (at point B on the curve), a drop in load and a
brittle breakdown of the block take place.

It is natural to assume that the fissures appearing during the splitting of ice samples could be caused by the normal
or shearing stresses.

Let us examine in more detail the process of insertion of
a flat punch into ice. In proportion to an increase of load on
the punch, the stresses in the ice mass increase and moreover
with particular intensity along the edge of the punch. In connection with this, the maximal tangential stresses developing
in the depth of the body at a certain distance from the punch
outline, reach a critical value much earlier than along the
symmetry axis. Therefore, slip planes develop in the material,
gradually progressing and then emerging jump-wide to the surface.
The phenomenon of the origination and development of cracks
(easily visible in the layer of a transparent ice sample) took
place several times during the testing process.

Starting from the time of splitting, the pressure distribution becomes more uniform. A further increase in the load
led to an increase in the tangential stresses under the punch
until the critical values were attained along the symmetry axis

of the specimen. After this, there took place the total breakdown of the material for a known depth, or overall plastic flow
began.

In this manner, one should consider that in case the
critical state is determined by the tangential stresses which
is valid for the square, round and "blunt" triangular punches.

At sharper outlines of the punch, the stress state becomes
complicated by the bulging effect of the intensified pressure
zone during the forcing of it into the body's interior [131].
This leads to the appearance of tensile stresses and to the subsequent breakdown of the flows. Since the value of the developing tensile stresses decreases rapidly at distance from the ice
apron blade, it is possible to have the appearance of fissures
of a limited length, causing only partial splitting of the floe.

Summing up, we can arrive at the following conclusions:

1. The experiments undertaken on determining the resistance
of ice to the penetration of various punches into it demonstrated
the significant influence of the punch shape.

2. In front of the punch, as a result of the volumetric
stressed state, a compacted zone is formed, the configuration of
which varies under various punches. With a blade of rectangular
shape, the zone has the form of a half-cylinder. For the triangular blades, the compacted zone becomes more elliptical, its
dimensions decrease and finally at a tapering angle of the ice
apron blade $2\ \alpha < 60°$, the zone is not formed.

3. The reason for such a variation in the form and dimensions of the compacted zone is found in the varying nature of
stressed state during the action of the "blunt" or "sharp" punches   on the material.

4. As a result of the action of blades of various shape
on the ice, the material's disruption sets in, proceeding in a
brittle manner, with the formation of fissures.

5. The force onto the flow during the introduction of a
support into it increases fairly gradually but unevenly. We find
a cessation in pressure rise and even a decrease in it at the
instant of crack's formation. Therefore, the load on a support
is of a vibrating nature.

6. The maximal resistance of ice to the penetration of
variously shaped punches into it is reflected in Table 25.

7. The effect of a punch's shape in case of the breakdown of large floes in general corresponds to the conclusions made in the theory of plasticity, as is obvious from Table 26.

Appreciable deviations occur only at small tapering angles, when the disruption under conditions of our experiments occurred chiefly in a brittle manner with the formation of cleavage.

8. The effect of the form of support in plane, as we observed, is quite significant. In the calculation of the ice pressure on a support cutting into a large floe, we suggest introducing the factor m for the support's form, taking its influence into account.

Table 25

Resistance of Ice, Percent, to the Introduction of Punches

test conditions	semicircular	Type of blade				rectangular
		triangular at 2 α				
		120°	90°	75°	60°	
With lateral compression (large floes)	86	82	65	54	41	100
Without lateral compression (small floes)	96	92	78	--	46	100
Average	92	86	71	54	44	100

Table 26

Effect of Punch's Form

2 α	100°	120°	90°	75°	60°	Remarks
P, %	100	82	65	54	41	based on tests
	100	81	71	66	61	based on (7.1)

If we assume that the forcing in of a square punch is typified by a form coefficient equalling 1.00, in the other cases in conformity with the test data and the conclusions of the plasticity theory, we can recommend the values for the form coefficient m listed in Table 27.

Table 27

Table of Values

Form of support's cutting face	Tapering angle 2α	m	$m = 0.85\sqrt{\sin\alpha}$
Semicircular	--	0.90	--
Triangular	120°	0.81	0.79
	90°	0.73	0.71
	75°	0.69	0.66
	60°	0.65	0.60

At sharper outlines of the punch, in our tests the samples, in breaking up, did not become crushed but became split. Therefore we can also extend the equation

$$m = 0.85\sqrt{\sin\alpha} \qquad (7.2)$$

coinciding well with the tests at angle 2α, equalling 120° and 90°, to the supports with a sharper configuration of the ice-cutting face.

THIRD PART

PRESSURE OF ICE ON STRUCTURAL ABUTMENTS (SUPPORTS)

Chapter VIII

CUTTING THROUGH LARGE FLOES BY ABUTMENT WITH VERTICAL
CUTTING EDGE

1. General Concepts

As has been shown above, the process of breakdown of an
ice sheet by the supports of structures proceeds variously, main
ly depending on the reserve of a floe's kinetic energy and its
strength. Since the travel velocity of large floes will not
fluctuate so intensively, a decisive role is played by the
dimensions of the ice field.

The small floes, having encountered a support, experience
local crushing at the point of impact and then, having exhausted
the available supply of kinetic energy, they stop, become turned
by the current and are borne to the adjoining span, if other
floes do not interfere with this. Floes of somewhat larger
dimensions usually become split while the large ice fields often
become disrupted by being cut through by their supports without
cracking.

We should also point out that in addition, the amount of
pressure of ice on the supports is in a complex relationship
with a number of additional factors, among which we can specifically include:

a) wind, which (changing the velocity of the floe's movement) can lead to another form of disruption;

b) the degree of ice concentration on the river surface,
which can make difficult the splitting of floes, promote their
stopping at the structural supports, and, on the other hand,
can transmit some of the kinetic energy from certain floes to
others in contact with the structural supports;

c) the form of support in plane, also exerting a considerable influence on the magnitude and direction of interaction forces;

d) the material and nature of processing the support surface, determining the more or less significant role of frictional forces;

e) the capability of the supports for a greater or lesser deformation; on which a part of the floe's energy can be expended

f) the eccentricity of impact, leading to a loss of part of the floe's energy in its rotation relative to the point of impact; and

g) the form of water surface in the region of support, causing deformation of the ice sheet and facilitating its disintegration.

Finally, as was shown above, the actual process of disrupting the ice cover is accompanied by a number of features, greatly complicating the theoretical review of the question.

One must never forget that in the period of spring ice passage, the properties of the ice cover are quite diversified, depending on the given stage of the development of the spring thawing processes. Even a short list of the factors determining the pressure of ice on the structural support indicates the great complexities of the problem and the desirability of undertaking a natural determination of value for the ice pressure.

However, since the difficulties originating in a natural determination of the value for the ice pressure have been appreciable until recent times (for details, see Chapter 17), the appearance of various theoretical methods permitting at least an approximate evaluation of the pressure value is natural.

We should mention the indisputable preeminence held in this field by the domestic and especially Soviet scientists. Here it is pertinent to list the names of L.F. Nikolai [40], M.A. Rynin [42], G.P. Perederiy [39], A.N. Komarovskiy [1], N.M. Shchapov [41], O.A. Dubakh [49], Ye. V. Bliznyak, as well as A.I. Gamayunov [46], B.V. Zylev [45], Ye. Palatonov [50], P.Z. Kuznetsov [47] and other researchers, having made various proposals on the determination of the value of ice pressure on structural supports. The studies made by the author [54,53,55, 58] are also devoted to this same question.

All the methods suggested until now can be divided into three groups:

a) the methods serving or determining the pressure of large ice fields which are being cut by the structural supports;

b) the methods taking into account the pressure of small
floes, stopping in front of a support after a partial breakdown
of the floe edge at the impact point; and

c) the methods reflecting the pressure of floes having
smaller dimensions, breaking up as a result of splitting.

Let us analyze in sequence all the groups of proposals
advanced, wherein we will initially have in mind the following
limitations:

1. The supports of the structures under consideration are
fairly massive, so that we can disregard the effect of their
deformations.

2. The nature of finishing the ice-cutting edges of a
support is such that the effect of frictional forces is insignificant.

3. The support's cutting edge is vertical and has a
tapered outline in plane.

4. The effect of wind is taken into consideration in selec
ting the calculated travel velocity of a flow (according to the
procedure discussed in Chapter 18). The question concerning the
effect of the deformations of a support, of the slope of the
ice-cutting edge to the horizon and the form of outline of a
support in plane is reviewed subsequently (Chapters 11-13).

2. Analysis of Calculation Procedures Previously Suggested

The suggestions on determining the pressure of ice on
structural supports in this case were made by N. A. Rynin [4],
having correctly considered that the pressure cannot be greater
than the forces capable of breaking up the flow. Later on, this
viewpoint was supported by G.P. Perederiy [39], N.M. Shchapov
[41], A.N. Komarovskiy [43], A.A. Surin [133], P.A. Kuznetsov
[47] and by many others. In addition, the same proposal was at
the basis of the Technical Specifications of the MPS on planning
the railway bridges (TUPM-47 and TUPM-56) and also GOST 3440-46
on determining the ice loads on structures [134].

In this context, the recommendation was usually made to
find the value of disruptive force based on the ultimate crushing
strength of ice according to the equation

$$R = FR_{c}\varkappa , \qquad (8.1)$$

where $F = b_{CM} h$ = the vertical projection of contact of support and ice floe; $R_{compres}$ = ultimate compressive strength of ice; b_{CM} = width of crushing area; and h = thickness of floe.

It is evident that pressure P reaches a maximum after the cutting of the support into the ice by a complete width b_o at

$$F = b_o h. \tag{8.2}$$

Equation (8.1) does not take into account such significant features of the disruption process as the phenomenon of local crushing, effect of load rate, or the influence of support's form in plane. Therefore, it is quite natural that it has been subjected to refinements. Thus, in 1933, A.N. Komarovskiy [1] suggested the acceptance into the calculation of only part of the support work, corresponding to its pressing into the ice for a depth of 7-15 cm (assuming that the splitting of the floe would take place after this). However, such an assumption is valid only for floes of small dimensions and is not acceptable for the condition of displacements and crushing of large ice floes.

In these cases, it is quite possible to have pressure on a girder part of the support as was observed repeatedly by the author [54] and also by A.S. Ol'mezov and G.S. Shpiro [69] under the actual conditions of supports' functioning. Nevertheless, as we have shown above, in reality the pressure of an ice pack almost never (with the exception of very weak ice) takes place over an entire perimeter of the ice-cutting edge, in connection with which the author [54] as early as 1938 suggested finding the calculated width of the contact area b_p based on the equation

$$b_p = kb_o$$

where k = coefficient of contact completeness (0.40 - 0.70).

Then a number of authors, whose studies have been discussed in more detail in Chapter 6 undertook extensive research into the mechanical properties of ice for establishing the calculated value of the ultimate crushing strength, $R_{compres}$. In this connection, the suggestion was made to take into account the loading rate and local crushing, since the appreciable importance of this factor was clarified. It turned out that the effect of local crushing is described well by the relationship:

$$R_{cм} = R_{cж} \sqrt[3]{\frac{B}{b_{cм}}} \leqslant 2{,}50\ R_{cж}. \tag{8.3}$$

In addition, in an analysis of the effect of the deformation rate, the author suggested computing the ultimate compressive strength of ice based on the relationship:

$$R_{cx} = \frac{R_1}{\sqrt[3]{v}} \quad (v > 0,1 \; \text{м/сек}). \tag{8.4}$$

Only after conducting the experiments on resistance of ice to the insertion of punches of varying form, the necessity and possibility of at least an approximate consideration of the support's form in plane by introducing a coefficient of the form

$$m < 1.00$$

was clarified.

3. Analysis of Effective Technical Specifications and Standards

Many of our suggestions (consideration of the load crushing phenomenon, form of support in plane, requirement for the regioning of pressure value) were embodied in the currently effective Technical Specifications for determining the ice loads on river hydraulic engineering structures and bridges (SN 76-59).

Pressure on a support in the direction of its axis is proposed to be determined based on the equation

$$R = mR_{compres} \; b_o h, \tag{8.5}$$

where $R_{compres}$ = ultimate shattering strength of ice where consideration of local crushing equalling 45-75 tons/m^2, depending on level of ice passage; and m = coefficient of support's (pier's) form.

The suggestion is made to increase the calculation values $R_{compres}$ by 2 times for the rivers opening during negative air temperatures and also for the rivers located north of Petrozavodsk, Kirov, Petropavlovsk, Novosibirsk, Ulan-Ude, Birobidzhan and Magadan.

It should be pointed out that SN 76-59 differs advantageously from GOST 3440-46 and TUPM-56 by virtue of a greater validity and a consideration of a number of new factors (local crushing, form of pier in plane). Nevertheless, it is possible and necessary to introduce certain corrections into them based on the following concepts.

a) Equation (8.5) provides the maximal possible ice pressure on a structural support, developing during the action of large ice fields on a solid pier. However, in many cases, the possible dimensions of ice fields preclude us from counting on such forces of interaction.

The floes of smaller sizes, having exhausted the reserve of kinetic energy in the disruption of their own edges, will stop or become split. In this connection, the forces of interaction will be less and can also be regulated by the Technical Specifications.

b) In the Technical Specifications, one should indicate the limits of application to Eq. (8.5) depending on the mass of floe and the velocity of its movement.

c) The division of the vast territory of the USSR with its quite diversified climatic conditions into only 2 zones is too approximate. The strength of the ice cover up to the period of breakup depends on the pattern of air temperatures, on the "intensity" of spring warming, on the features of ice structure and on other factors.

Therefore, it is desirable to give a more differentiated division of the USSR territory into zones based on the conditions of spring breakup of rivers.

d) We should examine the effect of the rate of floe's movement on the ultimate strength of the ice pack and refine its calculation values.

In connection with this, we discuss below a substantiation of the methods used in determining the forces of ice pressure on a support during its cutting into a large floe.

4. Mathematical Relationships

Let us trace the variation in the value of ice pressure on a support of pointed form in plane in proportion to its penetration into a large ice field.

In the process of penetration of the pier, the width b_{cm} of crushing area gradually increases and finally reaches a maximum value equalling the width b_0 of the support, after which the ice pressure can be assumed constant.

In this context, let us examine two stages of the phenomenon.

A. First Stage

It is obvious that at each time moment, ice pressure on a support can be assumed to equal

$$P = F_{CM} \ R_{CM} \ m, \qquad (8.6)$$

where $F_{CM} = kb_{CM} \ h$ = contact area of floe and support; R_{CM} = ultimate crushing strength of ice which for the large floes will always be assumed to equal $R_{CM} = 2.5 \ R_{compres}$; and m = coefficient of pier's form.

Replacing F_{CM} and R_{CM} by their values, we will write Eq. (8.6) in the form:

$$P = 2.5mkhb_{CM} \ R_{C \varkappa} \ .$$

Also having in mind that the ultimate crushing strength $R_{compres}$ of ice is linked with the velocity v of floes' motion by Eq. (8.4), we can also propose another expression for P, specifically:

$$P = \frac{2,5R_1 mkb_{c_M}h}{\sqrt[3]{v}}, \qquad (8.7)$$

where R_1 equals ultimate compressive strength of ice at floe velocity of 1 m/sec.

To avoid confusion, let us recall that Eq. (8,7) was obtained by processing the experimental data corresponding to $v \geqslant 0.10$ m/sec and is therefore valid only within these limits. At lower values of v, the flow of ice under a load is possible, wherein the increment in the R-value is limited by the limit of plastic ice flow.

B. Second Stage

After the support has cut into the ice for its complete width (b_0), the increase in the pressure stops, and P reaches the maximum value where consideration of (8.4):

$$P_{max} = \frac{2,5mkb_0hR_1}{\sqrt[3]{v}} \qquad (8.8)$$

or,this expression can be written:

$$P = K_p b_o h,$$

(8.9)

where

$$K_p = \frac{2,5 m k R_{cж}}{\sqrt[3]{v}}.$$

Equation (8.9) is the most general formula and as we shall see, it takes into account: a) climatic features of place of erecting the structure, $R_{compres}$; b) form of pier in plane m and its dimensions b_o; c) the incompleteness of contact of support and floe k; d) the velocity of floe movement v; and e) the phenomenon of local crushing (coefficient 2.5).

5. Comparison with Test Data

The equations (8.6-8.9) which have been derived can be compared with the results of our experiments in determining the resistance of river ice to the forcing of punches of various shape into it.

It was indicated previously that the ultimate compressive strength of ice is associated with the deformation rate by the relationship:

$$R_{cж} = \frac{3 \cdot 10}{\sqrt[3]{S}} = \frac{3 \cdot 10}{\sqrt[3]{\dfrac{w}{H}}} \quad Kg/cm^2$$

wehre w = rate of machine's crossarm travel, in cm/sec; and H = height in cm of sample.

In our tests, k = 1, b_o = 2.5 cm, h = 7.0 cm, H = 12 cm, w = 1.5 cm/min = 0.025 cm/sec.

Then the deformation rate:

$$S = w/H = 0.0021 \ sec^{-1},$$

and the ultimate compressive strength:

$$R_{cж} = \frac{3 \cdot 10}{\sqrt[3]{S}} = 24,6 \quad Kg/cm^2.$$

In this case, a calculation based on Eq. (8.8) yields:

$$P = 2.5 \text{ m} \cdot 1 \cdot 2; 5 \cdot 7 \cdot 24.6 = 1080 \text{ m} \qquad (8.10)$$

A comparison of the experimental data with the calculations based on Eq. (8.10) is given in Table 28.

Table 28

Results of Experiment

Number of tests	angle of pointing of edge	m	P, Kg actual	P, Kg (after 8.10)	average % of deviation
67	180°	1.00	1053	1080	
68	180°	1.00	1189	1080	
59	180°	1.00	862	1080	−10.3
70	180°	1.00	1189	1080	
71	180°	1.00	1174	1080	
58	120°	0.80	1023	865	
60	120°	0.80	894	865	− 7.2
59	120°	0.80	862	865	
53	90°	0.70	731	760	
54	90°	0.70	829	760	
55	90°	0.70	698	760	− 5.5
56	90°	0.70	698	760	
9	75°	0.65	843	703	
10	75°	0.65	698	703	− 8.0
32	semicircular	0.95	992	1025	
33	"	0.95	1008	1025	− 2.5

As we will observe, the discrepancy of the test data as compared with the calculations based on Eq. (8.8) is relatively slight (7.3% on an average) which testifies to the admissibility of utilizing the relationships which have been derived.

6. Recommended Method for Determining Maximum Ice Pressure

Equations (8.6-8.9) provide a possibility of finding the ice pressure during disruption of a large ice field by a support (pier).

For practical application, it is more convenient to utilize an equation of the type:

$$P = K_p b_o h. \tag{8.11}$$

It is obvious that the specific calculated pressure can be found with the formula

$$K_p = \frac{2.5 m k R_{c \text{ж}}}{\sqrt[3]{v}} \tag{8.12}$$

with the following values of calculation parameters:

a) The ultimate compressive strength of ice (at velocity of floes' movement of 1 m/sec) can be assumed to equal: for the rivers of the North and Siberia, $R_{compres} = 50$ tons/m^2, and for the rivers of the European sector of the Soviet Union, $R_{compres} = 30$ tons/m^2.

b) The coefficient k of incomplete contact should be assumed in the limits from 0.4 to 0.7. It is natural to assume that with a decrease in velocity of floes' movement and of support width, the k-value increases (Table 29).

Table 29

Recommended Magnitudes of the k-value

Width of support, m	Velocity of floe movement, m/sec		
	0.5	1.0	2.0
3 - 5	0.70	0.60	0.50
6 - 8	0.60	0.50	0.40

In conclusion, let us formulate a method of determining the calculated pressure in the following approach.

The maximum pressure which is being absorbed by a solid pier during its cutting of a large ice field can be calculated with the equation

$$P = K_p b_o h, \tag{8.11}$$

where K_p = specific calculated pressure established according to Table 30.

The suggested table is based on extensive test material and also on the utilization of a number of natural observations and theoretical concepts discussed in detail above.

As we shall observe, the suggested standards make it possible to take into account on a more selective basis the form of the support in plane, the velocity of floe movement, and also the climatic features of the region where the structure is located.

Table 30

Table of K_p-Values

Form of cutting edge of support in plane	Times of ice passage		
	displacement	total ice passage	
	v=0.5 m/sec	v=1 m/sec	v=2 m/sec
Semicircular	59	41	27
Triangular at tapering angle 2 = 120°	54	37	24
90°	49	33	22
75°	45	32	21
60°	43	30	20
45°	40	28	18

Remarks. 1. The K_p-values are given for the conditions of the ice passage in the middle and lower reaches of rivers in the European sector of the USSR, flowing southward (at $R_{compres} = 30$ tons/m^2. 2. For the rivers of the North and Siberia flowing northward, K_p can increase up to 1.7 times. 3. For the supports with a width of 6-8 m and more, the calculated values can be decreased additionally by 1.3 times. 4. For the rivers located under other climatic conditions, K_p can be calculated directly according to Eq. (8.12).

7. Limits of Applying the Technique under Study

It is obvious that the process analyzed above of the introduction of a support (for its entire width) into ice can take place only at a sufficient reserve of kinetic energy T at the floe being disrupted, which (energy) is expended chiefly in the work of the resistance (drag) forces of the ice cover to cutting of the support into it, $T_{разр}$ (for the depth x), depending on the dimensions of floe, velocity of its movement, strength of ice in the spring period, configuration of ice edge and finally the form of support in plane.

Toward the end of the period under consideration, the floe velocity decreases to v_1 m/sec, which determines the final energy reserve T_k.

It is obvious that the equation of energy balance can be written:

$$T_и \approx T_{разр} + T_k \qquad (8.13)$$

Proceeding from Eq. (8.13), we can find the amount of energy required for the cutting of the support into the ice for the given depth x_0.

Obviously,

$$T_{разр} = \int_0^{x_0} P\, dx. \qquad (8.14)$$

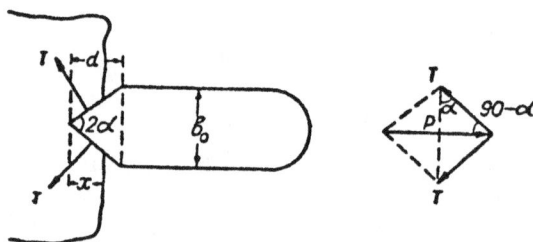

Fig. 34. On the Process of Penetration of a Pier into an Ice Field.

Prior to the integration, let us express P through x. For the most common supports with a triangular edge (Fig. 34), with consideration of (8.8), we have

$$P = mkhR_{cж} \sqrt[3]{Bb_{cм}^2} = 1{,}6mkhR_{cж}\sqrt[3]{Bx^2 \operatorname{tg}^2 \alpha}. \qquad (8.15)$$

We can now establish the value $T_{разр}$, which for $x \angle d$ proves to equal

$$\qquad (8.16)$$
$$T_{разр} = mkhR_{cж}\sqrt[3]{Bx_0^5 \operatorname{tg}^2 x}$$

or

$$T_{разр} = 0{,}6P_к x_0, \qquad (8.17)$$

where P_k = pressure on support at end of considered period for x_o = d. The energy reserve T_d required for the cutting of the support into the ice for its (the support's) entire width (b_o) will be obtained on the basis of (8.17) by the substitution of x_o equals d, after which we obviously have,

$$T_d = 0.6 \, P_{max} \, d = 1.5 m k h b_o R_c \text{\tiny ※} \, d, \qquad (8.18)$$

where P_{max} = maximal ice pressure on support being determined according to (8.8).

Thus, the pressure on the support reaches the maximal possible value being established on the basis of (8.8) or (8.11) only if the reserve of the floe's kinetic energy reaches a value greater than:

$$T_d = 1.5 m k b_o R_{cm} d,$$

The area of the floe sufficient for the appearance of the maximally possible pressure can be determined from the inequation:

$$\frac{\Omega h v_{\pi}^2 \gamma}{2g} > 1{,}5 m k h b_o R_{c \text{\tiny ※}} d. \qquad (8.19)$$

At the usual b_o-values (3-5 m), k (0.6-0.7), $R_{compres}$ (50-100 tons/m^2), the tapering angle (2 α = 75°-90°) m (0.7-0.9) and v_{π} (0.5-2 m/sec), such an energy is possessed by floes with an area ranging from 1 to 10,000 m^2. As we shall observe, even the relatively small floes (of the order of 30·30 to 100·100 m) during the time of full-scale ice passage are capable of causing the penetration of a support for its entire width and of exerting the maximal possible pressure on the support.

In this manner, the case under review has great practical significance.

8. Certain Conclusions

1. A large floe moving at velocity v_{π} m/sec can be stopped or cut through by structural supports located in its path.

2. The force of ice pressure on the support at any moment can be found with the equation:

$$P = m k b_{cm} h R_{c \text{\tiny ※}} \sqrt[3]{\frac{B}{b_{cm}}}. \qquad (8.6)$$

3. After the cutting of the support through the total width b_o, the pressure reaches a maximum value equalling

$$P_{max} = 2.5mkb_ohR_c \text{*}.$$ (8.8)

4. For practical purposes, it is more convenient to utilize a dependence of the form:

$$P_{max} = K_pb_oh.$$ (8.11)

5. The maximal pressure on the support determined according to (8.8) or (8.11) develops only if the floe has an initial kinetic energy supply greater than

$$T_H > 0.6P_{max}d,$$

where d = length of tapered part of ice apron blade.

The area of floe sufficient for the appearance of maximum pressure can be established from the inequation

$$\Omega > \frac{16mkb_0^2R_{cж}}{v_я^2 \operatorname{tg}\alpha}.$$

Usually such a force is possessed by a floe with an area ranging from 1 to 10 thous. m^2 and more.

6. In the period of full-scale ice passage at complete coverage of the river surface with ice, it is quite possible to have the combined action of individual floes which, exerting pressure on each other, seemingly combine their efforts for the disruption of the ice sheet. Under these conditions, it is possible to have the development of a critical value of pressure from the floes with a smaller area, of the order of 1-3 thous. m^2.

7. The suggested method refines the recommendations of SN 76-59, takes into account certain additional factors (completeness of contact, velocity of floe) and evaluates more selectively the climatic features of the region.

Chapter IX

CASES OF STOPPAGE OF SMALL ICE FLOES

1. Original Concepts

As we shall observe, even the relatively small floes have, during the full-scale ice passage, a supply of kinetic energy sufficient for the cutting of the support (pier) into the ice for its total width b_o. Moreover, the force of ice pressure on the pier reaches its maximum and can be calculated on the basis of Eq. (8.11).

If the supply of the floe's kinetic energy is less than the required minimum, the floe can be stopped even before the support manages to cut its entire width into the ice. It is quite natural that in this case the maximum force of ice pressure P on the support will be less than P_{max}.

Therefore, the determination of the pressure force developing during stoppage of small floes has less practical interest than the problem reviewed earlier. Undoubtedly, for practical purposes on many large and medium-sized rivers in the USSR, it is possible to have the appearance of floes capable of causing a maximal pressure of ice on the supports, particularly in the initial period of ice passage.

However, during the full-scale ice passage (or movement of scattered ice), the dimensions of floes decrease and in this case the determination of the ice pressure on a support is of a certain interest. In addition, a determination of the force developing during the stoppage of a small floe also has significance in designing the structural supports, which are erected on small rivers or in reservoirs, characterized by a slow current velocity.

2.   State of the Question

In determining the ice pressure on the supports of bridges and hydraulic engineering structures, in the case under review, we have suggested a number of methods developed almost exclusively by the Russian and Soviet scientists.

As early as 1897, L.P. Nikolai [40] proposed a method for determining the pressure of ice on bridge supports, based on an analogy with the phenomenon of the central impact of free inelastic bodies. However, L.F. Nikolai's method did not gain much popularity chiefly owing to a number of provisional assumptions placed at the basis of the calculation (assumption of reduced mass of support and of span structures, neglect of deformations of the base, support and floe).

Later on, A.A. Dubakh [49], A.N. Komarovskiy [1] and V.T. Bovin [5] suggested finding the ice pressure, proceeding from the theorem of pulses and momentum based on the familiar equation

$$P = mv/t. \qquad (9.1)$$

Let us note that in 1939, a similar formula was recommended by the plan of the Technical Specifications of the Scientific-Research Institute of Hydraulic Engineering, i.e. TS SRIHE [167].

As we shall observe, for the application of Eq. (9.1), it is necessary to know the stopping time of floe (t), which however can never be found without studying the resistance of ice to fracturing under local crushing conditions. At the same time, A.N. Komarovskiy considered it necessary to assume approximately for all cases t = 1 sec, while in the Technical Specifications, on planning the hydraulic engineering structure, the SRIHE recommended assuming the stopping time at $\Omega = 100$ m^2 t = 1 sec, $\Omega = 10\ 00$ m^2 t = 5 sec.

Such recommendations do not take into account the actual pattern of the phenomenon.

In 1933, N.M. Schapov [41] suggested determining the maximal ice pressure on a support, proceeding from the assumption to the effect that kinetic energy reserve of the floe is spent chiefly in the work of breaking down the edge and he also recommended the method of finding the stopping time of the floe. In 1938, we [54] examined according to N.M. Shchapov's technique, the cases of impact of a floe by one of its corners of projection of triangular form in plane (diagram in Fig. 35). In this con-

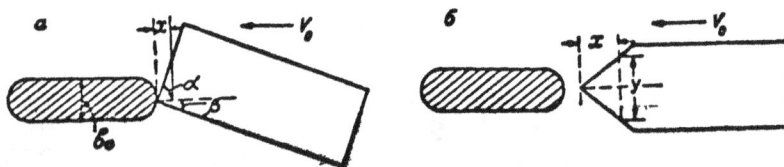

Fig. 35. Possible Positions of Floes at Time of Impact.

nection, we proposed the calculation methods and equations and specifically for the case of impact, schematically reflected in Fig. 35b, the relationship

$$P = vh \sqrt{\frac{2LB\gamma R \operatorname{tg} \alpha}{g}} = 0{,}43vh \sqrt{\Omega R_{cx} \operatorname{tg} \alpha}. \qquad (9.2)$$

In addition, the utilization of N.M. Shchapov's technique provided the possibility of demonstrating that the stopping time of a floe depends on the dimensions, strength and velocity of floe movement, form of support in plane and, for floes with an area up to 200 m^2, it comprises a range of 0.08 to 0.44 sec, which is appreciably less than 1 sec. Finally, an approximate method was suggested for finding the stopping time of a small floe according to the formula

$$t = 2x_0/v, \qquad (9.3)$$

being confirmed well by the more precise calculations based on N.M. Shchapov's procedure. Finally, the suggestion was made to utilize Shchapov's method with the introduction of a coefficient of incompleteness of support's contact in floe. In this manner, the proposals made by the TS SRIHE and A.N. Komarovskiy require correction.

In 1939, P.A. Kuznetsov [47] suggested a further development of Shchapov's method by taking into account the additional work spent in turning the floe around the point of contact with the structure, on the elastic deformations of ice and deformation of the structure.

The expression for the force of ice pressure in this case naturally became greatly complicated and could be written in the form:

$$P = 0{,}43vh \sqrt{\frac{\Omega}{\sin^2 \varphi \left(\frac{1}{nE} + 2ah \right) + \frac{\cos^2 \varphi}{\operatorname{tg}^2 \varphi} \left(\frac{0{,}91l^2}{nEh^2} + 8ah \right) + \frac{1}{R_{cx}}}}, \qquad (9.4)$$

where, in addition to the previously reproduced notations: E = elastic modulus of ice; A = coefficient of structural pliability; φ = angle between direction of floe motion and front (face) of structure; and n = percentage of floe width participating in its elastic deformations.

In reference to P.A. Kuznetsov's proposals, we can indicate that the effort toward a more complete inclusion of the various factors was conducted by him without considering the practical weight of any given corrections. As was indicated later by Kuznetsov and A.I. Gamayunov [46], the work spent on turning and the elastic deformations of a flow comprises less than 2% of the work for crushing and the consideration of the eccentricity of impact and elastic deformations of ice does not have practical meaning. Along with this, Kuznetsov completely overlooked the phenomenon of local crushing and the effect of velocity of floe motion, bearing much more significantly on the calculation results. In addition, in the derivation of Eq. (9.4), certain inaccuracies were permitted, as was correctly indicated by B.V. Zylev. It should be indicated that in 1946, Kuznetsov [51] already found it possible to simplify his known formula and to take into account only the deformation of a structure and the angle of the floe's approach to the structure.

Equation (9.4) assumed the form

$$P = 0{,}43vh \sqrt{\frac{\Omega}{\frac{\cos^2 \varphi \Omega}{\lg^2 \varphi n E h^2} + \frac{1}{R_{c.\kappa}}}} = cvh^2 \sqrt{\frac{\Omega}{\mu\Omega + \lambda h^2}} \qquad (9.5)$$

and was written into GOST 3440-46 for the determination of ice loads on hydraulic engineering structures. In 1946, Ye.V. Platonov [50], considering the phenomenon of the floe's encounter with a support arrived at the equation

$$P = 0{,}43vh \sqrt{\Omega R_{c\kappa} \lg \alpha}, \qquad (9.6)$$

which is quite identical to our Eq. (9.2) published as early as 1938. However, while Eq. (9.2) described the impact of the triangular projection of a floe on a support, considering the impact of a flat floe on the support, it is necessary to take into account the phenomenon of local crushing and the velocity of floe movement, which was not done by Ye. V. Platonov.

In 1948, B.V. Zylev subjected the various methods of calculation to a detailed analysis, and for the case of supports with a vertical cutting face, he suggested the use of Eq. (9.6). For the supports with a semicircular shape of cutting face, B.V. Zylev developed, based on N.M. Shchapov's technique, a mathematical formula of the type

$$P = rhb_{CM} R_{compres} , \qquad (9.7)$$

where r = radius of support's cutting face in plane; b_{CM} = width of crushing area (at r = 1).

Unfortunately, in Zylev's recommendation, he also overlooked the effect of the $R_{compres}$-value of the local crushing phenomenon and the velocity of floe's movement.

Summing up, we conclude that notwithstanding the extensive work performed by a number of researchers, we must consider a further review of the problem necessary.

3. Bases of Proposed Calculation Procedure

The proposed calculation method proceeds from the following assumptions.

1. The floe approaching the structure has a known reserve T_H of kinetic energy being spent subsequently mainly in the work of breaking up the ice and deforming the structure.

2. Since it is insignificant, we will disregard the effort spent in elastic deformations of ice and turning of floe during eccentric impact.

3. We have in mind that the reserve of kinetic energy of a floe is insufficient for cutting of the pier into the ice for the total width, i.e.

$$T_H < T_d$$

at $T_d = 0.6 R_{max}d$.

4. At the basis of the calculation, we have placed the concepts of Shchapov, supplemented however by a consideration of the local crushing phenomenon, the rate of floe movement, and the weakening of the ice up to the time of spring ice passage (ice-out).

5. The form of the support's cutting face in plane is assumed triangular or semicircular.

6. We have examined above only the solid supports, for which the allowance for deformations is insignificant. A method for considering the deformations of supports of a lightened type will be presented later.

Let us proceed to a discussion of the proposed method.

4. Case of Triangular Outline of Abutment's Cutting Face (Edge)

A floe with a dimension of L · B · h, moving at velocity v_H m/sec has a kinetic energy reserve T_H, which is spent chiefly in the work of breaking up the ice. Pressure of ice on the support at any moment (Fig. 26) can be established with the relationship

$$P = F_{c_M} R_{compres}m, \qquad (9.8)$$

wherein, as was shown above, the ultimate crushing strength of ice cannot be greater than 2.5 $R_{c\,\varkappa}$.

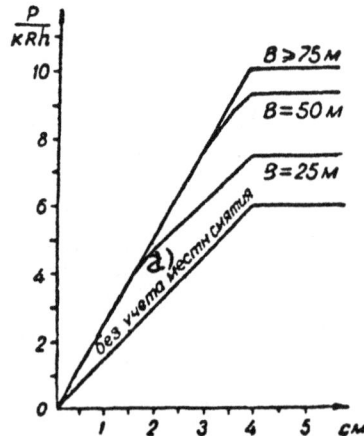

Fig. 36. Graph Showing the Increase in Pressure of Ice in Proportion to Penetration of Support into Floe. Key: a) without consideration of local crushing.

Obviously, if the floe has a slight reserve of kinetic energy, the support will be capable of entering the ice only for a slight depth.

Let us present the solution of the mathematical relationships.

Let us consider a floe with an energy reserve T $<$ T_d, moving with a velocity

$$v < \sqrt{\frac{R_{c\varkappa}m}{18\,tg\,a}}.$$

Since the width of crushing area (b_{c_M}) for the supports with a triangular cutting face equals

$$b_{c_M} = 2 \ x \ tan \ a \ , \ while \ R_{compres} = 2.5 R_{compres}$$

the pressure force can be established with the equation

$$P_1 = 5 \ m \ R_{compres} \ hx \ tan \ a \ , \qquad (9.9)$$

-132-

The energy reserve T required for penetration to the depth x is found from the expression:

$$T = \int_0^x P_1\, dx = 2.5m\, R_{cж}hx^2\, \mathrm{tg}\,\alpha, \qquad (9.10)$$

from which the depth of support's penetration into ice

$$x = 0.14v \sqrt{\frac{\Omega}{m\, \mathrm{tg}\,\alpha R_{cж}}}. \qquad (9.11)$$

Substituting the value x from (9.11) into Eq. (9.9), we find:

$$P = 0.68th\, \sqrt{\Omega R_{cж}m\, \mathrm{tg}\,\alpha}. \qquad (9.12)$$

Equation (9.12) is a final one for the case under consideration. Obviously, we can utilize it only until the inequation occurs:

$$\frac{B}{b_{cм}} = \frac{B}{2x_1\, \mathrm{tg}\,\alpha} \leqslant 15.6,$$

from which the critical depth of support penetration into the ice:

$$x_1 \leqslant \frac{B}{31.2\, \mathrm{tg}\,\alpha}. \qquad (9.13)$$

The necessary reserve of kinetic energy can be found from Eq. (9.10) and proves to equal (for square floes at $\Omega = B^2$):

$$T_1 = \frac{\Omega h R_{cж}m}{380\, \mathrm{tg}\,\alpha}.$$

It is interesting to note that, since

$$T_1 \leqslant \frac{\Omega h v^2 \gamma}{2g}, \qquad (9.14)$$

from a comparison of the T_1-values the requirement follows:

$$v < \sqrt{\frac{R_{cж}m}{18\, \mathrm{tg}\,\alpha}}. \qquad (9.15)$$

In this manner, the pressure from all of the floes (stopped by the support), moving with a velocity less than (9.5) can be calculated with Eq. (9.12). Of course, such a conclusion is valid only for the floes hitting against just one structural support. In case of impact of a floe against several structural supports, one can easily introduce the appropriate corrections into Eqs. (9.9), (9.12)

5. Case of the Semicircular Profile Cutting Face

For the case of a semicircular configuration of support, it is natural to apply the same pattern of arguments as was used in the case of a triangular form (outline). This problem was reviewed by N.M. Shchapov as early as 1933 [41]. It is obvious that at any moment:

$$b_{CM} = 2 \sqrt{x(2r-x}$$

and hence, the pressure force

$$P = 2R_{c*}h\sqrt{x(2r-x)}. \tag{9.16}$$

Equating the work of force P to the kinetic energy reserve of the flow, we can obtain:

$$\frac{Mv^2}{2} = \int_0^{x_0} P\,dx = Rh\left[\frac{R_{c*}r^2}{2} - (r-x_0)\sqrt{x_0(2r-x_0)} - r^2 \arcsin\frac{r-x_0}{r}\right].$$

Solving the derived equation for x_0 by trial-and-error, we can find b_{CM}, and then based on (9.16), we can find the pressure force P. As we shall see, this technique is inconvenient for practical use, which obviously instigated B.V. Zylev in 1948 to propose another solution to the problem, based on the following concepts [163].

In the pressing in a support, the total work spent in crushing the ice equals

$$FhR = mv^2/2, \tag{9.17}$$

where F = the area of segment based on the system in Fig. 37.

Having found from (9.17) the value F/V^2, it is easy to find the central angle (corresponding to r = 1), and then also the width of crushing area $\overline{AC}$, also corresponding to r = 1.

Then, finally

$$P = \overline{AC}rhR_{compres}.$$

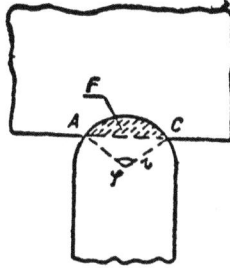

Fig. 37. On a Calculation of Ice Pressure on A Support
of Semicircular Configuration, According to
B. V. Zylev.

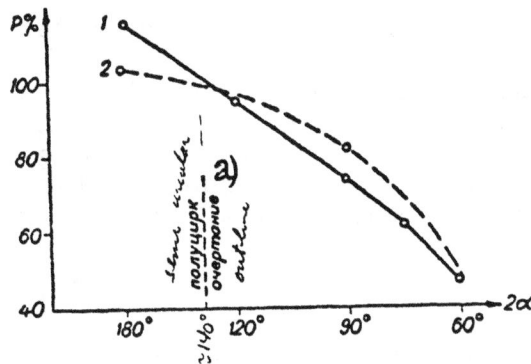

Fig. 38. Dependence of the Coefficient m on the Face
of the Pier, According to Experiments of the Author.
Key: a) semicircular outline.

In this manner, the problem according th Zylev is already
solved more simply. It can readily be seen that neither Shchapov
nor Zylev took into account the local crushing phenomenon or the
effect of loading rate, which would have complicated the problem
considerably. Therefore, we have suggested a different solution
to the problem.

In Chapter 7, it has been shown that at the insertion of
punches of various shape into the ice, a compacted zone is formed
(having a circular outline), through which the action of the pier
on the ice sheet is transmitted. The tests showed that the piers
of semicircular and blunt triangular outline are essentially
equivalent to each other.

Considering the test results, we can assume the semicircular punch as equivalent to the triangular one, with a tapering
angle of around 140° (Fig. 38).

Let us validate this assumption by an example. N.M. Shchapov [41] examined the impact of a floe with a size of 25X8X0.7 m (its travel speed was 2.6 m/sec) against a support with a width of 4m with a semicircular outline of cutting face. The ultimate compressive strength of ice was assumed to be 200 tons/m^2. The width of crushing area $b_{C M}$, found by cancellation proved to equal 1.96 m, and the pressure force:

$$P = 200 \cdot 0.7 \cdot 1.96 = 273 \ m$$

B.V. Zylev solved this sample example by using his method, and he derived:

$$b_{CM} = 0.955 \ m,$$

$$P = b_{CM} \ rhR = 0.955 \cdot 2 \cdot 0.7 \cdot 200 = 268 \ tons$$

Let us apply the assumption advanced by us (concerning the equivalency of a support of semicircular outline of a triangular support at $2\alpha = 140°$) to the solution of the given problem. Since N.M. Shchapov and B.V. Zylev solved the problem without consideration of local crushing, it was natural to utilize Eq. (9.2) or (9.6).

At $\alpha = 70°$, we get

$$P = 0{,}435 \ vh \sqrt{\Omega R_{cx} \ tg \ \alpha} = 0{,}435 \cdot 0{,}7 \cdot 2{,}6 \sqrt{200 \cdot 200 \cdot 2{,}75} = 263 \ m.$$

As we shall see, the results are quite close to N.M. Shchapov's and B.V. Zylev's calculations which confirms the admissibility of our assumptions.

Thus the pressure developing at stoppage of floe at a semicircular support can be found with Eq. (9.12) at $\alpha = 70°$ and m = 90. Let us write the converted expressions for the value of pressure P and maximum depth x_0 of pier's penetration into the ice:

$$P = 0{,}68 \ vh \sqrt{\Omega m \ tg \ \alpha R_{cx}} = 1{,}07 vh \sqrt{\Omega R_{cx}},$$

$$x_0 = 0{,}087 \ v \sqrt{\frac{\Omega}{R_{cx}}}.$$

(9.18)

6. Brief Conclusions

1. If the floe has a slight reserve of kinetic energy, it can stop at the structures before the pressure on the support reaches its maximal value P_{max}.

2. Such a phenomenon develops if the kinetic energy reserve of the flow satisfies the inequation

$$T < 0.6P_{max}d, \tag{9.18}$$

where d = length of support's tapered part.

3. Determining the ice pressure during stoppage of floe at a solid support, when

$$T < T_d \text{ and } v < \sqrt{\frac{R_{сж}m}{18 \, \mathrm{tg} \, \alpha}},$$

The ultimate crushing strength of ice R_{CM} can be assumed constant and equal to

$$R_{CM} = 2.5 \, R\text{compres}.$$

4. The maximal pressure developing at the end of the process of the floe's stopping

$$P = 0{,}68vh\sqrt{\Omega R_{сжm}\,\mathrm{tg}\,\alpha} \leqslant K_p b_0 h. \tag{9.12}$$

5. For a support of semicircular outline, we recommend using the same formula, assuming in it $\alpha = 70°$, after which it acquires the form

$$P = 1.07 \, vh\sqrt{\Omega \, R_{c*}} \tag{9.18}$$

6. Everything which has been said pertains to the solid supports, the deformations of which can be overlooked. In other cases, it is possible to consider the supports' deformations based on the procedure discussed below.

Chapter X

CASES OF SPLITTING THE ICE FLOES WITH ABUTMENTS

1. Original Concepts

The observations of the operation of the ice aprons to the supports, discussed in detail above, have demonstrated that in many cases after the introduction of a support into ice, the complete or partial splitting of the floe takes place. At this time, we usually note the appearance of a "leading" crack ahead of the support, which developing in proportion to the support's penetration into the ice, leads to the complete splitting of the floe into two parts. Let us point out that the leading crack sometimes had a length up to 5-30m and most often was directed by 10-30° to the side from the support's axis. The zigzag form of this crack in the plane which had been observed at times testified to the possibility of the opening of old cracks during impact; these cracks appear abundantly in spring during the fluctuation in water levels and displacements of ice.

It is also necessary to point out that the splitting of a floe usually takes place before the support is cut into the ice by a depth x, equal to or greater than the length of tapered part d (Fig. 34). This is quite natural if we keep in mind that after the introduction of the support for the depth d, the increase in the pressure of ice ceases. Therefore, if the floe could not be split at penetration depth $x \leqq d$, there are no reasons for expecting its splitting later on, and only when almost the entire floe has been cut through by the support, the splitting of the remaining section is possible.

2. Survey of Previously Suggested Calculation Methods

A.N. Komarovskiy [1], discussing the process of the introduction of a support into ice, gave the following expression for the pressure of ice (used by Kirkham [162] in designing the bridge piers across the Missouri R.):

$$P = \frac{R_p (L-x) h}{\sin \alpha \cdot \cos \alpha} .$$

(10.1)

It can easily be seen that Eq. (10.1) was obtained from the assumptions concerning the uniform distribution of tensile stresses along the section and concerning the inevitability of a simultaneous total splitting of the floe into two parts, with which we can never agree, at least for the floes of more or less considerable proportions. Therefore, Eq. (10.1) cannot be recommended for application.

In 1938, we suggested [54] a method for finding the pressure of ice on a pier, developing during the splitting of a floe. It was postulated that the floe with the dimensions L, B, h moving with the velocity v_Π m/sec, struck a pier with force P, wherein there took place the introduction of the ice apron blade for a depth S_0 and a splitting over the length Sm. Then the movement of the floe continued with a reduced velocity of v_1/m/sec.

-138-

Determining the loss of kinetic energy from the floe after encountering the support and equating its work and force P being spent for breaking the floe, we can obtain the length of crack

$$S = \sqrt{\frac{LR\gamma\,(v_{_{\Lambda}}^{2} - v_{1}^{2})}{gR_{p}\sin 2\alpha}} \qquad (10.2)$$

and a pressure force

$$P = hSR_p\sin 2\,\alpha\ . \qquad (10.3)$$

In Eq. (10.3), we took into account the partial splitting of the floe but the employment of this equation became complicated owing to the necessity of finding the velocity of the floe's movement after impact (v_1), which as is known depends itself on the amount of pressure. In addition, it is obviously necessary to take into account the share of the floe's kinetic energy spent in crushing the ice in the initial period of penetration. In connection with this, in the present report, we suggest a somewhat different solution to the problem.

In 1946, Ye. V. Platonov [50] recommended determining pressure on a support during the splitting of floes, proceeding from the assumption that during the introduction of a support into the ice for a depth y, on the ice field there would act two forces T, each of which equals

$$T = hyR_p.$$

Assuming the longitudinal pressure during splitting as equal

$$P = T\sin 2\,\alpha$$

and substituting

$$y = vt = 0{,}216\,v\,\sqrt{\frac{\Omega}{R_{cx}\,\mathrm{tg}\,\alpha}}\ ,$$

Ye. V. Platonov obtained

$$P = 0{,}216\,vh\,\sqrt{\frac{\Omega}{\mathrm{tg}\,\alpha}}\cdot\frac{R_{p}}{\sqrt{R_{cx}}}\sin 2\alpha. \qquad (10.4)$$

The last expression of the values (recommended by Ye.V. Platonov) R_p = 100 tons/m^2 and $R_{compres}$ = 350 tons/m^2 acquires the form:

$$P = 1{,}16vh\,\sqrt{\frac{\Omega}{\mathrm{tg}\,\alpha}}\sin 2\alpha. \qquad (10.5)$$

We are unable to agree with Platonov's ideas for the following reasons:

a) in the initial period, the support is introduced into the ice during the phenomena of plastic compression of ice and the amount of pressure on the support is determined by the plastic flow limit of ice rather than by the ultimate shearing strength R_p of ice;

b) based on the following data, we can judge the nature of the dependence of force during the splitting based on Platonov's formulas upon the tapering angle of the ice apron edge:

$$2\alpha = 60° \qquad 90° \qquad 120°$$

$$\frac{\sin 2\alpha}{\sqrt{tg\,\alpha}} = 1.16 \qquad 1.00 \qquad 0.66$$

As we shall observe, it turns out that the "blunter" outlines of the ice apron blade facilitate the splitting of a floe, which contradicts the natural observations;

c) Platonov's conclusions are made without allowance for the phenomenon of local crushing.

3. Bases of Suggested Calculation Method

The following tenets are at the bases of the suggested calculation method:

1. The floe under consideration, moving with velocity v_{π} , has a known kinetic energy reserve, which is being expended in encountering the support, chiefly for the work involved in breakdown.

2. After the meeting of the floe with the ice apron blade, there occurs the introduction of support into the ice, accompanied by a reduction in the velocity of the floe.

3. Ahead of the support's cutting face, a complex stressed state develops; its analysis leads to the conclusion concerning the significant role of the shearing stresses in the breakdown process, in any case, for the support having the semicircular and "blunt" triangular outlines of cutting face.

4. In case of the sharper outlines of the starling (or for the very small floes), the stressed state ahead of the support is complicated by the expanding action of the blade, which leads to the appearance of tensile stresses and to the subsequent breakdown of the floe.

5. Splitting of a floe is possible, in any event, only prior to the time when the support has cut into the ice for its total width, which is quite natural if we consider that after penetration of the support for the depth $x \gg d$, the increase in ice pressure ceases.

6. Therefore, the force of pressure developing under the splitting of the floe is always less than the force acting on a support during its cutting into a large floe.

Let us proceed to the derivation of the mathematical equations.

4. Diagram of Effective Forces

General Remarks

Let us consider a support with a triangular shape of cutting face, having penetrated into a floe with the dimensions $L \cdot B \cdot h$ for depth x and sustaining the pressure from the ice sheet (pack).

From our observations of the functioning of starlings in bridge supports, we can conclude that floes of relatively small dimensions are usually subjected to splitting. As was shown above, in order for the support to be unable to penetrate the ice for the total width b_o, the floe should have the area (8.19)

$$\Omega = \frac{16 m k b_0^2 R_{c\varkappa}}{v_{\textrm{л}}^2 \, \text{tg} \, \alpha} .$$ (10.6)

For the usual conditions ($2\alpha = 90°$, m = 70; k = 70; $R_{compres}$ = 50 tons/m^2), the maximal dimensions of the cleaving floes are small, which permits us to be limited to a discussion of the disruption of small floes, in which the crack emerges to the lateral or end face (edge) (Fig. 39). In accordance with the test data, let us assume that the crack's shape is rectilinear while its direction is deflected by the angle β from the support's axis.

External Forces

Let us consider a support with a triangular shape of cutting face, having penetrated for the depth x into a floe and sustaining the pressure P of the ice sheet. The forces of interaction developing between the support and floe can always be reduced to two standard forces P and to two frictional forces F, indicated in Fig. 39.

From the equilibrium conditions, we find

$$P - 2T \sin \alpha - 2F \cos \alpha = 0$$

Since the frictional force $F = fT$, accordingly

$$T = \frac{P}{2(\sin \alpha + f \cos \alpha)} \qquad (10.7)$$

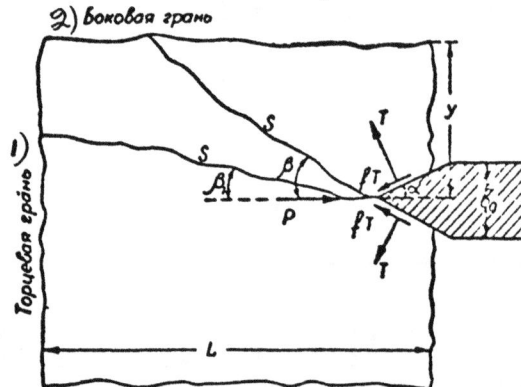

Fig. 39.   Diagram Showing Cracking of Floe and Developing Forces.   Key: 1) end face; and 2) lateral face.

It is then easy to find the component N, perpendicular to the pier's axis:

$$N = \frac{P(1 - \tan \alpha)}{2(\tan \alpha + f)} \qquad (10.8)$$

Thus, the pier penetrating the flow exerts the pressure T on it, which can be divided into the normal pressure N and longitudinal $P_1 = P/2$ (Fig. 40).

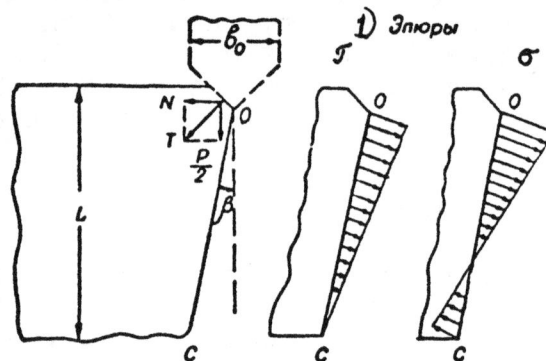

Fig. 40. Splitting of Small Floes.   On the right, sketches of tangential τ and normal σ stresses.   Key: 1) diagrams.

Let us note that the frictional coefficient between the pier's body and the ice sheet is generally low. In the practice of planning the icebreakers, the frictional factor is assumed to equal 0.01-0.05 [143]. The tests conducted by V.A.Arnol'd-Alyab'yev on the ice on the Neva. R., yielded for "moist" friction (according to painted steel) the values f = 0.10-0.15. Taking into account that in the survey of the problem concerning the splitting of floes, it is necessary for us to make various assumptions simplifying the problem, we consider it possible to assume f = 0, considering that the error developing thereby will be slight and will fall in the accuracy limits of our calculations.

Then, instead of (10.7) and (10.8), we find

$$T = \frac{P}{2 \sin \alpha} ;$$

$$N = \frac{P}{2 \operatorname{tg} \alpha} ; \qquad (10.9)$$

$$P_1 = \frac{P}{2} .$$

Distribution of Stresses

As a result of the action of a pier on an ice floe, in the latter, a complex stressed state develops, finally leading to the splitting of the floe into two parts.

Splitting the floe into two parts in a section posssibly running along the direction of crack OC (see Fig. 40), we can correctly assume the appearance in this section both of tangential and normal stresses, distributed according to some unknown law. Taking into account the studies conducted by N.A. Dmitreva [135] in respect to the distribution of stresses in the shearing plane of wood, and also by M.V. Plaksin [136] on the distribution of stresses during the splitting of wood with a wedge and bearing in mind the complete absence of experimental data for ice, as a first approximation, we adopt the diagrams of τ and σ indicated on the right in Fig. 40.

As we observed previously, in the floes of somewhat considerable proportions, the lateral compressive effect renders the possibility of disruption from cleavage as of little probability. In this case (and also during the encounter of a floe with "blunt" outlines of a cutting face), the disruptions from shearing stresses is most likely. On the other hand, in the cases of small floes (and of sharp outlines of the ice cutting face), disruption from normal expanding stresses are more likely.

Of course, such assumptions simplify and schematize the phenomenon, the actual physical pattern of which is incomparably more complex, and in connection with this can be regarded as merely a first approximation to a solution of the problem.

5. Splitting the Floes of Small Dimensions

Let us discuss the problem, after having postulated that in this case the splitting of a floe is caused by the shearing stresses originating in the plane of the crack's appearance with length S_1. Along some possible direction of crack S_1, let us cut the floe into two parts and consider the equilibrium of one of them. Signifying by θ the angle between direction of crack S_1 and direction of force T, we find (Fig. 41)

$$T(\cos \theta + f \sin \theta) = hSR_c = A \qquad (10.10)$$

where A = the resistance of ice sheet to splitting for length; and R_c = average splitting stress along line OC. It is obvious that the consideration of average shearing stress along length OC is possible only for small floes. Noting that

$$\theta = 90° - (\alpha + \beta),$$

and utilizing (10.9), we find

$$P = \frac{2R_c hS_1 (\sin \alpha + f \cos \alpha)}{\sin (\alpha + \beta) + f \cos (\alpha + \beta)}. \qquad (10.11)$$

Disregarding the frictional forces, we have a value of force sufficient for splitting the floe for the length S,

$$P = \frac{2R_c hS_1 \sin \alpha}{\sin (\alpha + \beta)}. \qquad (10.12)$$

Obviously, Eq. (10.12) is valid for the partial splitting of a floe for the length S. In case of total splitting of the floe into two parts, it is possible to have the emergence of a crack on the lateral face or even on the end face opposite the pier, according to the diagram in Fig. 39. However, the emergence of a crack on the end face of the floe **opposite the pier is less likely**, since the length of floe is usually greater than width, the length of crack (even in case of the meeting of the pier with a square floe) is minimum during splitting in the direction of the lateral face.

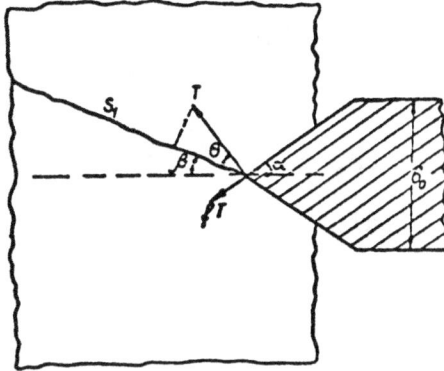

Fig. 41.   On Calculating the Forces Developing During
the Splitting of a Floe.

Obviously, during lateral splitting,

$$S_1 = Y/\sin \beta$$

and pressure P can be calculated with the formula

$$P = \frac{2R_c hY \sin \alpha}{\sin \beta \cdot \sin (\alpha + \beta)} .$$

(10.13)

Obviously, the splitting will take place at such an angle
(β) at which, at minimum P-value, the maximum cleavage stresses.

Investigating Eq. (10.12), we find the most likely value
of angle

$$\beta = (90° - \alpha)$$

and the working value

$$P = 2R_c h \sin \alpha .$$

(10.14)

For splitting the floe sideways (10.13), we accordingly
find

$$\beta = (90° - \alpha /2)$$

and

$$P = 4 R_c hY \tan \alpha /2$$

(10.15)

In case of central impact (Y = B/2), the equation is
simplified to:

$$P = 2 R_c Bh \tan \alpha /2$$

(10.16)

Thus, we have obtained relationships providing a possibility of calculating the amount of force necessary for breaking up small floes under the action of shearing stresses.

6. Splitting the Small Floes

In the case of small ice floes or piers of very sharp configuration, it is possible to have disruption from the normal tensile stresses developing in the plane of a fissure.

Having reviewed this question in detail [145], we can offer the following equations in this case:

$$\beta = \alpha/2,$$

$$P = nLhR_p \tag{10.17}$$

and the coefficient of pier's form n, varying in relationship to the angle of its tapering:

$2\alpha =$	60°	70°	80°	100°	120°C
n =	0.25	0.29	0.33	0.43	0.53

A detailed justification has not been presented since this case is usually not a calculation problem and in connection with this, does not represent significant practical importance.

7. Comparison With Experimental Data

For the confirmation of the relationships obtained, we undertook experiments in determining the resistance of ice to the insertion of punches of various shape into it, described in detail in Chapter 7. Let us compare the equations derived with the test data.

Taking into account that under actual conditions, the width of a floe is most often less than its length, we should consider in case of the encounter of large floes with the piers, the splitting (cleavage) will occur along the line of least resistance, i.e. sideways, according to the diagram shown in Fig. 39. Therefore, for the given case the force should be established on the basis of Eq. (10.16), which can be represented in the form:

$$P = \sigma_1 \tan \alpha/2, \tag{10.18}$$

where $\sigma_1 = 2 R_c Bh$ is a constant value for the given conditions. In this manner, the forces required for the splitting of a floe by piers of various shape are established with the value tan $\alpha/2$

A comparison of the test and calculation data is given in Table 31.

As we observed, the coincidence of the calculation and test data is quite satisfactory.

Table 31

Comparison With Test Data

Value	2 α				Remark
	60°	75°	90°	120°	
the same,%	$0.26\ \sigma_1$ 46	$0.32\ \sigma_1$ 58	$0.41\ \sigma_1$ 71	$0.58\ \sigma_1$ 100	Based on (10.18)
P experimental, %	49	65	78	100	

8. Brief Conclusions

Summing up, we can form the following conclusions:

1. During the introduction of a pier with a sharpened cutting face into a floe, the force of interaction can reach a value adequate for splitting the floe.

2. Only the small floes, in which in any case the supply of kinetic energy:

$$T < 0.6\ P_{max}\ d$$

can be subjected to cleavage.

3. The force necessary for splitting a floe is always less than P_{max}.

4. The splitting of a floe can occur in three ways:

   a) by splitting of the side section;
   b) partial splitting of length S_1; and
   c) splitting lengthwise into two parts.

5. The splitting takes place in the direction of development of maximum shearing or tensile stresses. The angle of a crack's deflection from the pier axis (β) can be established with the relationships:

in case a) $\beta = 90° - \alpha/2$;

in case b) $\beta = 90° - \alpha$;

in case c) $\beta = \alpha/2$.

6. The force capable of accomplishing the splitting of a floe equals:

in case a) $P = 2 R_{ck}Bh \tan \alpha/2$;

in case b) $P = 2R_{ck}hS \sin \alpha$; and

in case c) $P = nR_pLh$,

where L, B, h equal the length, width and thickness of floe; R_{ck}, R_p equals the ultimate shearing and tensile limit of ice; and n equals the coefficient calculated on the basis of Eq. (10.17)

7. The equations are confirmed by the experiments undertaken by us on the insertion of punches of various shapes into river ice.

In the calculations, it can be assumed (at ice temperature of around 0°, at orientation of pressure perpendicularly to the crystals' axis under the conditions of Siberian rivers):

a) the average disruptive stress during the cleavage of small floes

$$R_{ck} = 20\text{-}25 \text{ tons/m}^2,$$

b) disruptive tensile stress:

$$R_p = 70\text{-}75 \text{ tons/m}^2.$$

8. Everything that has been said pertains to the cleavage only of small floes. The cleavage of large floes is always partial, with the formation of a leading crack with length S_1.

Chapter XI

EFFECT OF SLOPE ANGLE OF ICE -CUTTING FACE

1. Introductory Remarks

For counteracting the danger of the formation of ice jams, resulting in damage or destruction to the structures, until recently, in the USSR use has been made of piers with a sloped ice-cutting face, finished in an appropriate manner, within the fluctuation limits of the ice passage levels. However, notwithstanding the fact that the use of a sloped face often leads to a considerable increase in the scale of operations and hence to an increase in cost and an extension of the construction schedules, the question concerning the methods of selection of slope angle of the ice-cutting face and even the question concerning the role of a sloped face cannot be considered sufficiently developed until recent times. Certain specialists consider the basic purpose of a sloped ice-cutting face to be a decrease in the pressure of the ice on the pier, while others theorize that such a decrease does not take place. Some consider the most dangerous time of ice passage to be the moment of the first thrust of ice, and they suggest building the sloped face within the limits in the fluctuations of displacement levels; other consider it necessary to build a sloped face only within the fluctuation limits of high ice-passage levels.

Some persons assume that in the bridges with small spans, one should not build an inclined face (owing to the danger of the floes' jamming between the lengthened supports); on the other hand, others consider that with small spans, the sloped faces are more desirable since they facilitate the destruction of ice. Therefore, the necessity for the further development of the question cannot be in doubt.

2. Role of Slope of Ice-Cutting Face

Various opinions have been expressed on the question of the division of the forces on the inclined face of an ice apron. For example, the Technical Specifications for the planning of bridge piers (TUPM) and the hydraulic engineering structures (GOST 3440-46), adopted the method of dividing the force of ice pressure on the inclined edge of the ice apron (starling).

-149-

The pressure force P is divided into components, one of which (P_1) causes the tendency of a floe to rise along the ice aprong edge while the other (N) acts directly on the structural pier. The force

$$N = P \sin \beta ,\qquad (11.1)$$

in its turn can be divided into the components:

$$\text{vertical} - P \sin \beta \cos \beta \qquad (11.2)$$

$$\text{and horizontal} - P \sin^2 \beta ,\qquad (11.3)$$

which were also taken into consideration in verifying the strength and stability of the pier.

Many specialists assume that the division of the force of ice pressure during encounter with the inclined face of the pier generally does not take place [50,71].

A.N. Komarovskiy [43] also noted that the dynamic pressure of ice on a support with the construction of a sloped face can be decreased only in exceptional cases at small dimensions of floes, not dangerous per se. It should be noted that both Ye.V.Platonov and A.N. Komarovskiy nevertheless recommend the application of sloped faces in rivers with particularly severe ice passage conditions or with rapid current velocities.

Let us indicate that the slight creeping of the floes onto an ice apron cannot serve as proof in favor of the absence of the division of the forces of ice pressure during encounter with a pier. This division takes place but the force tending to raise the floe of the ice apron is slight; in connection with this, one should not expect a considerable rise. In addition, owing to the relatively low shearing and breaking strength of ice, a large part of the floe breaks away (under the effect of its own weight), from the part climbing onto the ice apron; possibly this creates the impression of the absence of the floe's creeping up. The fact of the successful operation of the sloped wooden ice aprons (which, being vertical, are easily cut off by ice) fully confirms these findings.

Thus, the sloping of the ice-cutting face favors a reduction in the horizontal component of the pressure, leads to the appearance of a vertical component and by the same token, facilitates the pier's functioning. However, the basic purpose of the sloped face of the starlings should not be considered the reduction in ice pressure or the decrease in stresses in the pier body but the promotion of destruction of the ice sheet. The ice-cutting face, apparently cutting the ice field from beneath, removes the disrupted material upward and thereby favors the destruction of the ice sheet, avoiding the possibility of the formation of ice jams.

It is sufficient to recall the successful operation of an icebreaker, creeping onto an ice pack and pressing it down with its own weight. There can be no doubt that if icebreakers had vertical edges on the bow part of the ship, their attempts to break through the ice fields many kilometers wide would be doomed to failure.

However, considering that the application of the inclined ice aprons leads to a considerable increase in the scale and cost of operations (and also delays the time of putting the structure into operation), we suggest that one should resort to the use of an inclined face only in extreme cases, each time after a careful justification.

3. Possible Calculation Systems

After the meeting of a floe with the sloped face of an ice apron, the crushing of its sharp edge takes place, accompanied by the appearance of the reaction of the pier, which can be divided into forces V and T, interconnected by the equation

$$V = T \cot \beta. \qquad (11.4)$$

Under the effect of forces V and T, the destruction of the floe is possible (Fig. 42):

a) from bending--in section 1-1;

b) from shearing--in section 2-2;

c) from crushing under the action of forces T or even of their resultant:

$$H = 2 T \sin \alpha .$$

Obviously, the disruption will occur in some way or other depending on which stresses reach the critical value first. It is possible to demonstrate that the disruption from shearing is quite likely, based on the following concepts. As is known, the ice cover on rivers and lakes is comprised of the accretion of vertically oriented crystallites, separated from each other by the finest interbeddings of mother solutions of salts, having separated during freezing. As a result of the lower melting temperature, the spring thawing of ice begins from these interbeddings, in connection with which the shearing resistance of ice is greatly facilitated upon the crystals. Let us emphasize that in the cutting of a large floe by a pier, there remains a narrow belt of disrupted ice with a width about equal to the width b_0 of the pier. Having examined Fig. 42, we can also conclude that the ice rests closely against the pier and hence, the disruption from shearing occurs in direct proximity to the structure.

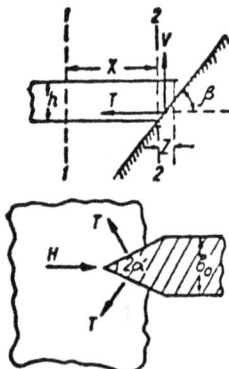

Fig. 42. Diagram Showing Interaction of a Floe With Inclined Ice Apron.

If the processes of spring thawing have not yet managed to develop in the necessary manner, we can have a disruption of the ice pack from bending. Let us point out that in this connection, the fracturing takes place in direct proximity to the pier at a distance not exceeding 3-6 thicknesses of the ice pack. The width of the channel cut by the pier in the ice field is somewhat greater than in the previous case, but differs only slightly from the pier's width. Destruction from crushing is possible only in the case of deep ice aprons when the vertical component is slight Therefore in this case, the floe will either stop at the pier or will break up from the crushing by force H.

4. Case of Breaking Up Large Floes by Shearing Force

Let us find the values for V and H at which the disruption of an ice pack by shearing occurs. Let us assume that the edge of the floe up to the time of disruption from shearing has been crushed for the length Z, as reflected in Fig. 42. Observations of the functioning of piers have shown that the disruption takes place directly at the support, roughly in the section 1-1, wherein the Z-value is slight.

We can then find:

a) the length of the shearing area:

$$b_{cp} = pk, \tag{11.5}$$

where p = perimeter of pier's cutting face, k = coefficient of contact looseness; and

b) the maximally possible vertical force exerted on one path of the pier

$$V_1 = b_{cp}hR_{cp}/2 \tag{11.6}$$

c) the maximally possible horizontal force

$$T = 1.1 \, V_1 \, \tan \beta_1, \tag{11.7}$$

where β_1 = inclination angle of lateral parts of pier's cutting face to the horizon, established from the expression:

$$\tan \beta_1 = \tan \beta/\sin \alpha \; ; \text{ and}$$

d) maximally possible horizontal force on the entire pier:

$$H_{max} = 2 \, T \sin \alpha = 2.2 \, V_1 \tan \beta . \tag{11.8}$$

Coefficient 1.1 is introduced in accordance with the suggestions made by B.V. Zylev for taking into account the friction between the pier and floe. Let us examine the specific cases of calculation.

Triangular Outline or Cutting Face

For piers with triangular outline of cutting face, we find the maximally possible length of shearing area from the expression:

$$b_{cp} = b_o \, k/\sin \alpha \tag{11.9}$$

and the design stresses for the entire pier, equalling:

$$V_{max} = 2V_1 = b_o k/\sin \alpha \quad hR_{cp} \tag{11.10}$$

and

$$H_{max} = 1.1b_0hR_{cp}k \ \tan \beta / \sin \alpha \ . \tag{11.11}$$

The value of ultimate shearing strength of ice R_{cp} was established for river ice at temperature of around 0°C and proved to equal during ice passage

$$R_{cp} = 20 - 40 \ m/m^2 \tag{11.12}$$

Let us point out that we experimented with ice which was still fairly strong when the bond between the crystallites was not appreciably weakened. The value for the coefficient of contact looseness of the pier and floe k depends on the current velocity, ice strength, dimensions and form of support in plane and can be assumed to fall within the limits ranging from 0.4 to 0.75 based on the data in Table 29 included in Chapter 8.

We can represent the mathematical equations in a more convenient form, namely:

$$H = K_{cp}b_0h, \tag{11.13}$$

where

$$K_{cp} = 1.1 \ R_{cp}k \ \tan \beta / \sin \alpha \ .$$

The K_{cp}-values for certain β, α, R_{cp} are grouped in Table 32.

Table 32

K_{cp}-Values

β \ 2α	60°	75°	90°	105°	120°	180°	Remark
45°	41	34	29	26	24	21	At
60°	71	59	51	45	41	36	$R_{cp}=25 \ m/m^2$
65°	89	73	63	56	51	44	
70°	113	93	80	72	65	57	k = 0.75
75°	154	126	109	97	89	77	
80°	233	192	165	148	135	118	

For other k-values, conversion is necessary.

Let us point out that the coefficient K_{cp} increases rapidly with an increase in angle β and at the limit (at $\beta = 90°$) it even reverts to infinity. Such a result is quite valid and natural since it indicates that in the case of steep ice aprons, the value of vertical component is slight and destruction occurs from crushing by force H and not from shearing by force V. Therefore, the calculation should be conducted based on the equation:

$$H_{cp} = K_{cp}b_0 h \leqslant K_p b_0 h \qquad (11.14)$$

Semicircular Configuration of Cutting Face

In this case

$$\begin{aligned} b_{cp} &= b_0 k/2 , \\ V_{cp} &= 1.57 h b_0 k R_{cp}, \\ H_{cp} &= K_{cp} b_0 h, \end{aligned} \qquad (11.15)$$

where

$$K_{cp} = 1.1 \, (1.57 \, \tan\beta \, R_{cp} k).$$

The K_{cp}-values are listed in Table 33.

As we observe, the sloped piers of semicircular outline are equivalent to the piers with triangular shape of cutting face at tapering angle of approximately $2 \alpha = 80°$.

Table 33

Values

β	45°	60°	65°	70°	75°	80°	Remark
K_{cp}	33	56	70	89	121	184	At $R_{cp}=25$ tons/m^2 k = 0.75

In this manner:

a) the maximum horizontal force transmitted by the flow to the inclined ice apron during its destruction from the shearing can be found in the formula

$$H_{cp} = K_{cp}b_o h$$

with the coefficient K_{cp} taken from Tables 33 and 32.

b) with steep ice aprons, the vertical pressure component (V) is small and disruption occurs from crushing by force H. Therefore the increase in K_{cp} should be limited by a certain value and assumed to be

$$H = K_{cp}b_o h \leqslant K_p b_o h$$

c) the maximum destructive effect on ice by means of shearing can be provided by the piers having the minimal perimeter. However, the poor streamlining of such piers can lead to undesirable results (washouts), in connection with which it is advantageous to use piers having a tapered outline of cutting face.

5. Destruction of Large Floe by Bending

The vertical component of ice pressure on a pier can lead to the disruption of an ice pack not only by shearing but also by breaking at a certain distance from the pier, as had already been indicated by L.F. Nikolai [40] as early as 1897, and later on by many other authors. The question of the operation of an ice field under the action of a vertical load applied to its edge was considered largely in connection with the operation of icebreakers, moreover this was done chiefly by Russian scientists [138, 73, 140]. In the question concerning the operation of the ice aprons around structural supports, a similar problem was first considered by B.V. Zylev [45], having applied the theory of the bending of an elastic isotropic semi-infinite plate on a resilient base (Fig. 43).

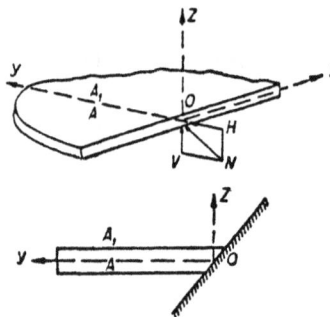

Fig.43. Formulation of Problem Adopted by B.V. Zylev.

Zylev arrived at the following conclusions:

1. The bending moment acting in the direction of axis (OX) reaches its maximum in the origin of coordinates and constitutes (at Poisson coefficient $\mu = 0.4$, bed coefficient $k_O = 1$ tons/m^3) around

$$(M_x)_{max} = 0.1 \quad V, \tag{11.16}$$

where V = the vertical component of pier's reaction.

2. The bending moment acting in direction of axis OY reaches its maximum at point A_1 with the coordinates

$$x = 0; \quad y = 1.14 \sqrt[4]{D} \tag{11.17}$$

and comprises around

$$(M_y)_{max} = 0.207 \, V,$$

which increases $(M_x)_{max}$ by almost two times.

Here D = the cylindrical stiffness of a plate equalling

$$D = \frac{Eh^3}{12 \, (1 - \mu^2)} \quad ,$$

h, E, μ = the thickness, elastic modulus and Poisson coefficient of river ice.

3. The moment of appearance at point A_1 with the coordinates

$$x = 0; \quad y = 1.14 \sqrt[4]{D}; \quad z = h/2$$

at normal stresses equalling the ultimate bending strength of ice (R), corresponds to a pressure equalling

$$V = \frac{R_{\text{и}}}{1,2 \frac{\eta_b}{h^2} + \frac{0,56 \, tg \, 3}{h \sqrt[4]{D}}} , \tag{11.18}$$

which Zylev suggests assuming for the maximally possible pressure. In Eq. (11.18)

$$\eta_b = \frac{\sqrt[4]{D}}{b_0} \left(sh \frac{b_0}{4\sqrt[4]{D}} - ch \frac{b_0}{4\sqrt[4]{D}} + 1 \right).$$

4. The maximal value of horizontal component of ice pressure equals

$$H_{max} = 1.1 \ V_{max} \ \tan \beta, \qquad (11.19)$$

where, with coefficient 1.1., we have roughly taken into account the role of the frictional forces between the pier and the floe.

The study conducted by Zylev has been devoted to an urgent problem, was conducted very carefully, and therefore is of interest.

The admitted approximation of the phenomenon leads to nonconformity between the theoretical conclusions and the experimental data. For example, based on theoretical assumptions, the disruption of the ice sheet should occur from the normal stresses caused by the bending moments (M_y), acting in the direction of axis (OY). However, as our observations indicate, the disruption begins from the appearance of cracks, oriented along the axis of the pier, i.e. associated with (M_x) rather than with (M_y).

A similar phenomenon is also noted at the Scientific-Research Sector of Navigation and Shipbuilding (SRSNS), emphasizing [72], [73] that during the breakdown of an ice pack by an icebreaker, there is initially separated a belt of ice bounded by cracks, and only then does the breaking of this belt occur with the formation of a transverse fissure (refer to Fig.4).

In addition, a calculation based on Eq. (11.18) indicates that disruption from bending can take place at a considerable distance from the pier, equalling from 12 to 19 thicknesses of the ice. As is known, the dimensions of ice chunks being broken off do not exceed 3-6 thicknesses of the ice pack and are often smaller (see Fig. 5, 6).

The studies conducted by Zylev are dedicated to the initial moment of a pier's encounter with a large floe and do not pertain to the cases when the pier has already cut into an ice field for a fairly considerable distance. In this connection, the force V will no longer be applied to the floe edge but somewhere in the central part of the drifting ice field, which indeed has already been weakened by a channel. Undoubtedly at this time the nature of the phenomenon will be different. In this manner, the research conducted by Zylev should be supplemented by a consideration of certain other factors.

6. Analysis of Effective Technical Specifications and Standards

In the Technical Specifications [146] for determining the ice loads on river hydraulic engineering structures and bridges (SN 76-59), the suggestion is made to determine the horizontal component of pressure on a pier with a sloped ice apron (along its axis) with the formula

$$H = P_{\text{и}} \, h^2 \tan \beta, \tag{11.20}$$

where $R_{\text{и}}$ = the ultimate bending strength of ice assumed to equal 0.7 R_{compres}; h = ice thickness; β = angle of cutting face's inclination to the horizon.

The recommendations given by SN 76-59 comprised a step forward as compared with GOST 3440-46 and TUPM-56 but nevertheless they inevitably cause certain arguments. In particular, in Eq. (11.20), no consideration is given to the width of pier: it is silently assumed that the pressures of an ice field on piers with a width of 3 and 10 meters will be identical.

In Eq. (11.20), no consideration is given to the influence of the pier's form in plane, which naturally fails to correspond to the physical nature of the phenomenon; the suggestion is made to assume an $R_{\text{и}}$-value equalling 0.7 R_{compres}, which constitutes from 31 to 105 tons/m^2.

We should assume that $R_{\text{и}} \approx R_{\text{compres}}$, without forgetting at this time that the value of R_{compres}, recommended by the TS, is given with allowance for the local crushing, i.e. is increased by 1.75 times. In our opinion, it is therefore possible to reduce the calculated values of $R_{\text{и}}$ by 1.25-1.5 times.

Equation (11.20) was obtained as a result of modelling the formula developed by B.V. Zylev.

The tests specially undertaken by us on the disruption of large ice fields by a vertical concentrated load [31] demonstrated that the solution obtained by Zylev reflects only the case of the static application of a vertical load during the absence of a horizontal one.

However, even here we have noted significant differences from the quantitative aspect.

The width of the disruption zone proved to be 1.5-2 times smaller than the calculated width and the bending differed appreciably from that calculated.

It appears that these discrepancies are explained by the excessive schematization of the phenomenon, the underestimation of the dynamic nature of the applied load and of the horizontal forces.

With respect to the utilization of the studies by Zylev for determining the dynamic action of ice on structures, we should take into account the features of the interaction process of the ice field and pier. At the encounter of an ice field with a pier, after the local crushing of the ice edge, radial cracks appear, running from the pier at a certain angle to its axis. Subsequently, concentric cracks appear, separating a series of ice consoles. Obviously, such a nonconformity in the theoretical findings by Zylev with the natural observations is explained by the effect of the horizontal component of pressure force, complicating the stressed state and causing a cleaving effect. In addition, the high rate of applying the load, the effect of anisotropy of the ice pack and the fluctuation in its thickness are of significance.

It should be noted that B.V. Zylev excluded from the review the horizontal component of pressure force which obviously also causes the direct breakdown of ice (cleavage) earlier than the vertical component causes the crumbling of the ice plate's edge.

7. Suggested Calculation Method

Original Concepts

Any mathematical method recommended for application in engineering practice should correspond, to the greatest possible extent, to the physical pattern of the phenomenon. In connection with this, let us first recall the process of demolishing a large ice field by a pier equipped with a sloped ice-cutting face.

After the encounter of a floe with a solid support having a cutting face with length l, the latter cuts into the ice for a more or less considerable distance equalling $n_0 l$ (at $n_0 < 1$), depending chiefly on the form of pier in plane and profile, ice strength and velocity of floe's movement. After cutting into the ice, the pier exerts resistance against the movement of the

-160-

ice field, acting on it with its lateral faces (1-2 (Fig.44) tending to raise part of the ice field. At this time, a crushing of the lower edge of the ice pack occurs for a certain depth z, in connection with which the reactive pressure N of pier on floe can be divided into the components V_1 and T, interrelated by the dependence

$$T = V_1 \tan \beta_1. \qquad (11.21)$$

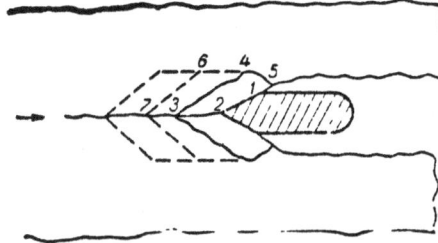

Fig. 44.   Diagram Showing Disruption of an Ice Sheet
by a Pier With a Slopped Face.

Let us emphasize that in this expression, β_1 equals the inclination angle of lateral faces of cutting edge to the horizon, connected with the inclination angle of the entire cutting edge by the relationship

$$\tan \beta_1 = \tan \beta / \sin \alpha . \qquad (11.22)$$

Then under the influence of forces V_1 and T, the breaking of the ice field occurs, accompanied by the appearance of at least 3 cracks (oriented along the pier's axis) separating two belts of ice, from which in turn the ice chunks 2-3-4-5 are broken. After this, pressure on the pier decreases, cracks 4-6 and 3-7 come together and the floe continues to move forward, not experiencing any serious counterthrust on the part of the pier until the latter once again comes in contact with the floe along the line 3-4 and the occurrence is repeated.   In this manner, the resultant reactive pressure of the pier on a floe reduces to two forces:

$$V = 2V_1$$

and

$$H = 2 T \sin \alpha .$$

Taking into account Eqs. (11.21), (11.22) and introducing (after B.V. Zylev) coefficient 1.1, taking into account the friction between the pier and floe, we find:

$$H = 2.2V_1 \tan \beta . \qquad (11.23)$$

Let us note that deep crushing usually does not take place Therefore forces V_1 and T are not applied along the center of the ice pack's depth but close to the lower ice surface (see Fig.42).

In this manner:

a) the process of breakdown of ice cover is characterized by great complexity and begins from the appearance of a number of cracks oriented along the pier's axis;

b) the disruption of the ice sheet has a periodic nature, wherein the pressure of ice on the pier reaches a maximum by the time of separation of an ice chunk, and then falls almost to zero; and

c) the role of the pier's form in plane and its width in the process of breaking down the ice pack is significant, just as in the case of the operation of icebreakers. Admiral S.O. Makarov and R.I. Runeberg [192] emphasized that the slope of an icebreaker's stem toward the horizon has a relatively low significance and they even suggest disregarding its influence, reckoning only with the average angle of attack, i.e. the inclination angle of the lateral faces of the vessel's bow to the horizon.

In connection with the considerable complexity of the phenomenon, we considered it possible to suggest an approximate calculation method, since a number of assumptions (which have to be made in utilizing the theory of bending of thin isotropic semi-infinite plates on an elastic base) are not justified in practice.

Working Formulas

Let us consider the solid support cutting into an ice floe with an inclined face of ice apron (with triangular outline of cutting face in plane) according to the system shown in Fig.44. As was explained previously, pressure on the edge of ice fields along the length 1-2 leads to the separation of an ice belt and then to the cleavage of an ice chunk along the line 3-4 at a certain distance y from the pier's face. Since the separation of the ice belt precedes its breakoff, we can assume that the maximum pressure on the ice develops during the fracture of the floe along line 3-4. In this connection, pressure is transmitted to the floe along line 1-2 according to length:

$$l_{1-2} = \frac{b_0 n_0}{2 \sin \alpha} \qquad (11.24)$$

while fracture occurs along line 3-4 with approximate length:

$$l_{3-4} = \frac{b_0 n_0}{2 \sin \alpha}.$$

Here n_0 = a coefficient less than unity, taking into account the pier's form in plane. Undoubtedly, with sharp outlines of the cutting face, fracture begins even before the pier has cut into the floe for its total width.

Since the width of the chunk of ice being separated is usually not more than 3-5 m, we can consider the belt 2-3-4-5 as a console, jammed in the section 3-4 and existing under the influence of forces V_1 and T. The actual weight of the console prior to encounter with the pier is compensated by the hydrostatic pressure and can be manifested only at emergence of the floe from the water. However, we should bear in mind that the deformation of an ice console prior to breakup is very slight [28,32]. Therefore we can overlook the effect of the actual weight.

Then the bending moment in section 3-4 will equal:

$$M = V_1 y - Td$$

or, with consideration of (11.21) and (11.22),

$$M = V_1 (y - d \tan \beta / \sin \alpha), \qquad (11.25)$$

The moment of resistance (drag torque)

$$w = \frac{h^2 l_{3-4}}{6},$$

and the maximum stress in section 3-4 up to the time of fracture can be computed with the equation:

$$R_x = \frac{6 V_1 \left(y - \frac{d \, tg \, \beta}{\sin \alpha} \right)}{h^2 l_{3-4}}, \qquad (11.26)$$

from which

$$V_1 = \frac{R_x h^2 l_{3-4}}{6 \left(y - \frac{d \, tg \, \beta}{\sin \alpha} \right)}. \qquad (11.27)$$

In order to use this expression for practical purposes, it is necessary to know the y-value, i.e. the distance from the pier to a section in which the fracture of the ice chunk takes place and also the arm of force d. Let us find them on the basis of many years' natural observations of the operation of the piers under bridges and hydraulic engineering structures.

We have at our disposal a number of photographs and movie films, from which it is possible to find the y-value as a function of thickness h of ice sheet. Analyzing them, we find the following relationships [5.8]: y = (2.5-6)h.

It is obvious that to the lesser y-value, there corresponds a greater value of force V_1 breaking the ice. Therefore, with a certain reserve, we can assume y = 3h.

With regard to the d-value, i.e. the arm of horizontal component H, it can obviously fluctuate within limits from 0 to 5 h, comprising 0.25 h on an average.

Thus, assuming

$$y = 3h, \quad d = 0{,}25h, \quad l_{3-4} = \frac{n_0 b_0}{2 \sin \alpha},$$

we find, instead of (11.27), the following expressions for determining the calculation forces:

the resultant maximum vertical pressure on a support:

$$V_{max} = \frac{0{,}66 R_\text{и} h^2 k \frac{b_0}{h} n_0}{12 \sin \alpha - \operatorname{tg} \beta} \; ; \tag{11.28}$$

the resultant maximum horizontal pressure on a support

$$H_{max} = \frac{0{,}73 R_\text{и} h^2 k \operatorname{tg} \beta \frac{b_0}{h} n_0}{12 \sin \alpha - \operatorname{tg} \beta} \tag{11.29}$$

or, more simply

$$H_{max} = R_\text{и} h^2 \operatorname{tg} \beta \frac{b_0}{h} S_0, \tag{11.30}$$

where

$$S_0 = \frac{0{,}73 \, n_0}{12 \sin \alpha - \operatorname{tg} \beta} \tag{11.31}$$

is the coefficient taking the pier's dimensions and form into account.

For finding coefficient $n_o S_o$, we proceed as follows. At encounter of a vertical support with a floe, the ice pressure value can be found with the relationship

$$P = 0{,}68 vh \sqrt{\Omega R_{\text{сж}} m \operatorname{tg} \alpha,} \qquad (9.12)$$

and hence the effect of form is reflected by the product

$$n = A \sqrt{m \tan \alpha} \; , \qquad (11.32)$$

where m = the coefficient of pier's form; 2α = the tapering angle; and A = the coefficient of proportionality.

In a first approximation, we assume that these tendencies are preserved also for the supports with sloped cutting faces.

Then taking into account that

$$m \approx 0.85 \sqrt{\sin \alpha} \; ,$$

we find the n_o-value equalling:

at 2α =	45°	60°	75°	90°	105°	120°
n_o =	0.47A	0.59A	0.71A	0.84A	0.99A	A

A comparison of the calculation results with the test data (see Part 6) provides the possibility of assuming A equalling 2. Then the values of coefficient S_o can be computed based on Eq. (11.31) for the prescribed actual conditions and are compiled in Table 34.

Table 34

S_o-Values

β \ 2α	45°	60°	75°	90°	120°
45°	0,20	0,17	0,16	0,16	0,15
60°	0,24	0,20	0,19	0,18	0,17
70°	0,38	0,27	0,23	0,21	0,19
75°	0,79	0,38	0,29	0,26	0,22

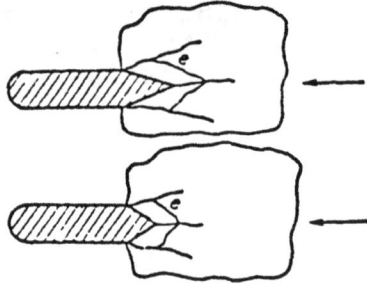

Fig. 45.   Diagram Showing Disruption of Ice Pack by
Piers with Varying Tapering Angle of Cutting Face.

As we shall observe, the effect of the tapering angle 2
of a pier is significantly reflected only at fairly small values
($2\alpha < 60°$) of it.

A certain increase in force H at very small tapering angles
can be explained by the fact that a sharper pier penetrates more
quickly into the ice field, and in connection with this, length
l of the ice chunks being broken off by the ice apron corres-
pondingly increases (Fig. 45).

Fig. 46.   Effect of Rounding of Pier's Cutting Face on
Nature of Ice Disruption.

In conclusion, let us note that one should take into account the rounding of a cutting face.

This rounding makes the edge more blunt, which leads to the appearance of additional leading cracks (Fig. 46).

In a first approximation, we suggest the following technique for taking this circumstance into account.

It is obvious that during the disruption of ice by piers with a sloped ice apron, of considerable importance is the perimeter along which the ice sheet comes in contact with the pier. Therefore the supports with a rounded cutting face can be regarded [147] as triangular with a reduced tapering angle:

$$2\alpha = 2x_1 + \frac{4r(40° - x_1)}{b_0},$$ (11.33)

while the semicircular ones are to be considered as triangular for allowance for the tapering 2α = 80° (Fig. 47).

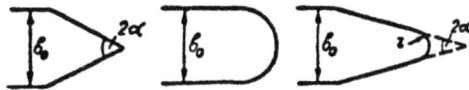

Fig. 47. Types of Piers Which Have Been Discussed.

We recommend that the value for the ultimate calculated bending strength be assumed at 50-65 tons/m^2 for the rivers in the North and Siberia and 30-40 tons/m^2 for the rivers in the European sector of the Soviet Union.

In Chapter 20, we have explained the procedure developed by us for establishing the actual ice pressure on structural supports, based on a study of the phenomenon's kinematics.

The use of this technique described in Chapter 22 for 18 cases on the Siberian rivers demonstrated the quite satisfactory coincidence of both the natural and calculated data.

Proceeding from this, we recommend Eq. (11.30) for practical application, since it

a) takes into account in more detail (than SN 76-59) the form of pier in plane (2 α, r), its relative width (b_0/h);

b) it estimates more correctly the actual strength of the ice pack during the ice passage period; and

c) it agrees quite satisfactorily with the natural data (at A = 2).

8. Numerical Examples

1. Find the maximum horizontal action of a large floe on a solid support during the period of thrust with the following original data:

$b_0 = 4m$; $R_{\text{и}} = 65$ tons/m^2; $R_{\text{compres}} = 65$ tons/m^2; $\beta = 60°$;

$h = 1$ m; $R_{cp} = 30$ tons/m^2; $2\alpha = 90°$ $k = 0.75$

$v = 0.5$ m/sec;

From the conditions of shearing based on (11.14), we have, using Table 32,

$$H_{cp} = K_{cp} b_0 h = 51 \cdot 4 \cdot 1.00 \quad 30/25 = 246 \text{ tons}$$

Since the table was compiled for $R_{cp} = 25$ tons/m^2, the tabular value K_{cp} (51 tons/m^2) is multiplied times the ratio of ultimate shearing strength.

From the condition of bending according to (11.30) and Table 34, we find

$$H = R_{\text{и}} h^2 \tan \beta \, b_0/h \, S_0 = 65 \cdot 1 \cdot 1.73 \cdot 4 \cdot 0.18 = 81 \text{ tons}$$

From the crushing conditions, we have:

$$K_p = \frac{2,5 R_{c\text{ж}} mk}{\sqrt[3]{v}} = \frac{2,5 \cdot 65 \cdot 0,73 \cdot 0,75}{\sqrt[3]{0,5}} = 75 \text{ tons/m}^2$$
$$H_p = 75 \cdot 4 \cdot 1 = 300 \text{ tons}$$

It is obvious that disruption takes place from bending and the maximum effect on the pier constitutes $H_{max} = 81$ tons. In this manner, the introduction of a sloped face decreased the pressure by $300/81 = 3.7$ times. Based on SN 76-59, in this case we would have had:

$$H = R_{\text{и}} \, h^2 \tan \beta = 105 \cdot 1 \cdot 1.57 = 181 \text{ tons}$$

2. Find the maximum horizontal effect of ice on a solid pier at retention of all conditions in example No. 1, but at a steeper ice apron with $\beta = 80°$.

Similarly to example No. 1, we have:

from the shearing condition:

$$H_{\text{cp}} = 165 \cdot 4 \cdot 1.00 \quad 30/25 = 792 \text{ tons.}$$

from the bending condition:

$$H_{\text{и}} = 65 \cdot 1^2 \cdot 5.67 \cdot 4 \cdot 0.43 = 630 \text{ tons.}$$

from the condition of shattering:

$$H_{\text{p}} = 75 \cdot 4 \cdot 1.00 = 300 \text{ tons}$$

The breakdown will occur from shattering and hence the use of such a steep ice apron (starling) does not make sense.

9. Brief Conclusions

1. The main purpose of the inclined face of an ice apron is to facilitate the disruption of the ice pack and to avoid the formation of ice jams.

2. The disruption of a floe during encounter with the sloped face of an ice apron can occur from shearing or bending for weak or strong ice, respectively.

3. The effective standards (SN 76-59) are more sophisticated than the old ones (GOST 3440-46 and TUPM-56), but overlook the width of pier and exaggerate the design limit of ice strength. In connection with this, their updating would be desirable.

4. We suggest the following equations for finding the maximum horizontal pressure force:

in the case of shearing:

$$H = 1.1 b_{\text{o}} h R_{\text{cp}} \quad k \tan \beta / \sin \alpha ; \qquad (11.13)$$

in the case of bending:

$$H = R_{\text{и}} \, h^2 \tan \beta \quad b_{\text{o}}/h \quad S_{\text{o}} \qquad (11.30)$$

5. The mathematical relationships for H during bending are compared with seven observations in nature and correspond to the latter, which permits us to recommend them for practical utilization.

6. With ice aprons which are too steep, disruption can occur from the crushing by force H before the vertical pressure V reaches a value adequate for disrupting the floe by means of shearing or bending.

In this manner, excessively steep starlings ($\beta > 75$-$80°$) do not yield the anticipated effect and therefore should not be utilized.

7. Taking into account the appreciable increase in volume and cost of operations, an inclined ice-cutting face should be used in solid piers only in extreme cases, and after a detailed justification.

8. Under the conditions of ice passage in the rivers of the European sector of the USSR, in all cases (and in the Asiatic sector, in many cases) we can apparently get by without application of sloped faces. Naturally, the wooden ice aprons should be provided with a sloped cutting face.

9. It is feasible to utilize more excessively the vertical faces of solid supports with sharper configurations in plane.

In many cases, such a solution will provide the possibility of replacing the costly sloped ice aprons.

Chapter XII

ALLOWANCE FOR DEFORMATIONS OF STRUCTURAL SUPPORTS

1. Abutments With Vertical Face (Edge)

In Chapters 8-10, we reviewed in detail the methods used in determining the ice pressure on solid supports with a vertical cutting face and we compared the proposed equations with the data derived from natural and experimental research. In this connection, we disregarded the deformations of solid supports based on their smallness; which this approach should substantiate.

However, in the application of piers capable of becoming more or less significantly deformed under the action of ice pressure, the consideration of the deformation can be appreciably reflected on the value of the developing forces of interaction.

A. Original Concepts

An ice floe encountering a structure in its path expends on it the available supply of kinetic energy mainly in breaking down the ice pack T_p, deformation of structure $T_{Я.с.}$, deformation of ice field $T_{Д.л}$ and turning the floe around the point of contact at eccentric impact $T_{вp}$.

If a floe, having exhausted the available supply of kinetic energy, stops before the pier has cut into the ice for the total width, the following equation is then valid

$$mv^2_{Л} /2 = T_p + T_{Я.с} + T_{Я.Л} + T_{Вp} \qquad (12.1)$$

where m and $v_{Л}$ = mass and initial velocity of floe's movement.

Examining the question regarding the effect of pier's deformation, we consider that:

a) an ice field, having a known supply of kinetic energy, initially expends part of it on the elastic deformations of ice and the deformation of the structure. If the remaining supply of kinetic energy is still adequate for penetration of the pier into the ice for the total width b_o, the maximal pressure can be found from the equation.

$$H_{max} = K_p b_o h, \qquad (8.1)$$

previously recommended in Chapter 8. In this case, the allowance for the work expended in deforming the structure and floe (and also for overcoming the frictional forces) does not exert an appreciable influence on the amount of maximal ice pressure, and therefore the consideration of the pier's deformation lacks practical significance here.

b) under actual conditions of the piers' functioning, the case is always possible when the impact will be centralized or almost so. Therefore, desiring the determine the maximum force of interaction between the structure and the floe, we can assume

$$T_{B?} = 0$$

c) based on the calculations by P.A. Kuznetsov [47] and A.I. Gamayunov [46], the consideration of elastic strains of floes changes the force of ice pressure slightly (by fractions of a percent). If in this connection we also take into consideration, as was noted correctly B. V. Zylev, that P.A. Kuznetsov exaggerated the work of the ice's deformation by three times, the possibility of also assuming that

$$T_{Я\,л} = 0$$

becomes obvious.

In this manner:

1. The pressure developing during the cutting by piers into large floes can be calculated with Eq. (8.11) without consideration of the other factors complicating a discussion of the question.

2. The pressure developing during the stopping of floes at the structural supports, can be found from the expression

$$mv_{л}^2 \ /2 = T_p + T_{Я.c} \tag{12.2}$$

derived from (12.1) with disregard of the work expended in the elastic strains and rotation of the floe.

B. Allowance for the Deformations of Structural Supports

In this manner, we can assume that a flow, having encountered a structure, utilizes the available supply of kinetic energy chiefly in the work of breaking up the ice and deforming the structure. Therefore, the more "pliable" the structure, the less will be the force transmitted to it. Signifying by 1/$\underline{a}$ the force necessary in order that the deformation of a structure along the line of force's effect would equal unity, let us express the deformation of structure in the form

$$\lambda_c = Pa$$

and work of deformation

$$T_{\text{д.c}} = \frac{P\lambda_c}{2} = \frac{P^2 a}{2}. \tag{12.3}$$

In connection with this, Eq. (12.2) acquires the form

$$\frac{\varkappa v_{\text{л}}^2}{2} = \int_0^{x_0} P\,dx + \frac{P^2 a}{2}. \tag{12.4}$$

Utilizing it, we can find an expression for P as a function of the stiffness $\underline{a}$ of the structure.

Let us limit ourselves to a review of piers with triangular outline of cutting face in plane, when the relationship

$$P = 5\,m\,R_{c\,M}\;hx\,\tan\,\alpha \tag{9.9}$$

takes place.

In this case

$$R_{c\,M} = 2.5\,R_{\text{compres}} = \text{const.}$$

From Eq. (12.4) with consideration of (9.9), we find

$$\frac{\varkappa v_{\text{л}}^2}{2} = \frac{Px_0}{2} + \frac{P^2 a}{2},$$

from which

$$P = 0{,}30\,vh\,\sqrt{\dfrac{\Omega}{ah + \dfrac{1}{5R_c m\,\tan\,\alpha}}}. \tag{12.5}$$

This relationship is a final one and can be utilized in determining the force of ice pressure with allowance for deformation of the pier.

At $\underline{a} = 0$, we find

$$P = 0{,}68vh\sqrt{\Omega R_{c\varkappa}m}\ \tan\ \alpha \tag{9.12}$$

i.e., Eq. (9.12) previously derived for the solid supports (piers)

2. Coefficient of Pliability

The coefficient of pliability of a pier for the system indicated in Fig. 48,

$$a = H^3/3EI, \tag{12.6}$$

where E = elastic modulus of pier's material; H = tentative depth of finishing; and I = inertial moment of transverse section of pier.

Let us point out that S.A. Bernshteyn [142], N.I. Bezukhov, P.A. Kuznetsov [47] and A.I. Gamayunov [46], discussing the problem of the impact of a floe with a set of piles, recommend assuming the inertial moment relative to the axis of the group as a unit. B.V. Zylev proceeds more cautiously, suggesting that we assume:

$$I = knI_1 \tag{12.7}$$

where k = a coefficient determined experimentally; n = number of piles in set; and I_1 = inertial moment of a single pile.

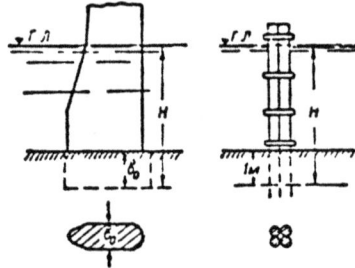

Fig. 48. On Calculation of Coefficient of Pier's
Pliability. Key: 1) ice level.

For the pile embankment or the pile grating, Kuznetsov recommends applying, in the case of rigid installation of pile heads in a ferroconcrete grating:

$$a = \frac{nl^3}{12E_1 J_1}, \tag{12.8}$$

-174-

while with a hinged installation of the pile heads, he recommends placement in a wooden grating:

$$a = n l^3/3E_1 I_1 , \qquad (12.9)$$

where l = free length of pile; n = number of piles in one row, perpendicular to the cordon; E_1 = elastic modulus of piles' material; and I_1 = inertial moment of individual pile's cross section.

Calculations based on Eqs. (12.6) - (12.9) yield the following values for the pliability coefficient a:

floating structures $\sim$ 0.01;
pile structures $\sim$ 0.001; and
solid structures $\sim$ 0.0001-0.000 001

Let us calculate what error we permit by disregarding the pier's deformations. The ratio of forces calculated with consideration of pier's deformations and without their consideration can be found from comparing Eqs. (12.5) and (9.12):

$$\lambda_1 = \frac{P_{(9\ 12)}}{P_{(12.5)}} = \sqrt{1 + 5mR_{cx}\, ahtg\, \alpha}. \qquad (12.10)$$

For the various tapering angles 2α of the pier in plane, we accordingly have:

2α = 75°   90°   105°   120°

m tan α = 0.50   0.70   1.04   1.39

and consequently, the maximum value of λ_1 equals

$$\lambda_1 = \sqrt{1 + 6.9R\ h.} \qquad (12.11)$$

Thus, the effect from considering the deformations of a structure can be determined on the basis of Eq. (12.11). At the values for (R, h) usually found, we can derive the λ_1-values in percentages, listed in Table 35. In this way, a consideration of the deformations of piers with a vertical cutting face has practical significance only for the floating (sometimes for the pile-type) structures. A calculation of solid structures can be conducted without allowance for the pier's deformations.

Table 35

λ_1-Values

a	1.10^{-8}	1.10^{-3}	$5\,10^{-3}$	$1\,10^{-4}$	1.10^{-5}	1.10^{-6}	$h, \text{м}$	$R_{\text{compres}},$ tons/m^2
% of increase in force	114,0 38,2	16,6 4,4	8,6 2,3	1,8 0,5	0,18 0,05	0,018 0,005	1 0,5	50 25

3. Abutments With Inclined Ice-Shearing Face

Let us examine the effect of deformation of a pier during the breakdown of an ice pack from shearing and crushing (separately).

Case of breakdown from shearing. The maximum horizontal pressure can be found from the equation

$$H = \frac{2,2hR_{\text{cp}}\,\text{tg}\,\beta}{\cos\alpha}x_0.$$

Substituting an expression for H into the energy balance equation:

$$\frac{\mu v_n^2}{2} = \int_0^{x_0} H\,dx + \frac{H^2 a}{2}$$

and solving the latter, we finally find

$$H = 0,30 v_n h \sqrt{\frac{\Omega}{ah + \frac{\cos\alpha}{2,2R_{\text{cp}}\text{tg}\beta}}}. \qquad (12.12)$$

Assuming a = 0, we find

$$H = 0,46 v_n h \sqrt{\frac{\Omega R_{\text{cp}}\,\text{tg}\,\beta}{\cos\alpha}}. \qquad (12.13)$$

In case of breakdown from crushing, we can utilize the suggestion made by B.V. Zylev [45], recommending the relationship having the form in our notations

$$H = 0{,}30 h v_{\text{a}} \sin^2\beta \sqrt{\dfrac{\Omega \varepsilon_1}{ah + \dfrac{1}{2k\,\text{tg}\,\alpha}}},$$

where

$$\varepsilon_1 = \dfrac{1}{1 + 3\cos^2\beta}.$$

This relationship is given without consideration of the local crushing phenomenon and under the assumption that the fracturing process is similar to simple compression. Therefore, it appeared useful to us to introduce certain corrections into it, having assumed

$$R_{c\,M} = 2.5 R_{\text{compres}},$$

after which the equation acquires the form

$$H = 0{,}30 v h \sin^2\beta \sqrt{\dfrac{\Omega \varepsilon_1}{ah + \dfrac{1}{5Rm\,\text{tg}\,\alpha}}}. \tag{12.15}$$

At a = 0, we find

$$H = 0{,}68 h v \sin^2\beta \sqrt{\Omega R m \varepsilon_1\,\text{tg}\,\alpha}. \tag{12.16}$$

The consideration of the pier's deformation in the given instance also leads to an increase in the calculated pressure H by a certain value.

In a comparison of Eqs. (12.15) and (12.16), we find

$$\lambda_{\text{M}} = \sqrt{1 + 5Rmah\,\text{tg}\,\alpha},$$

which at 2α = 120° acquires the form:

$$\lambda_{\text{M}} = \sqrt{1 + 6{,}95 Rah}, \tag{12.17}$$

very close to the expression which was analyzed previously. Consequently, also in the case of the presence of a sloped ice-cutting face, the effect of the piers' deformation should be taken into account only with a pliability coefficient of the structure a > 0.001.

Chapter XIII

EFFECT OF ABUTMENT'S FORM IN PLANE ON AMOUNT OF ICE PRESSURE

In Chapter 7, we showed the significant effect of a pier's form on the value of forces of interaction developing between a structure with a vertical face and a large ice field. In the subsequent chapters, we derived some equations for determining the amount of ice pressure, including the characteristics of the pier's form in plane. In this chapter, we present a brief analysis of the effect of a pier's form in plane and we also introduce certain concepts on an efficient configuration of a pier (support).

1. Case of Cutting Through Large Floes

In cutting through large floes by a pier with a vertical cutting face, one observes a crushing of the ice and its crowding out, mainly to one side, but in part upward and downward. It is known that if with blunt outlines of ice apron, the ice particles are displaced initially along the pier's axis and then are moved to one side, with sharp outlines of the ice apron's edge, the material as it were is scattered in a lateral direction; undoubtedly, this should lead to a decrease in the necessary force of ice pressure. The experiments undertaken by us confirmed this completely.

A.N. Komarovskiy [1] surmised that the pier's form does not influence the amount of pressure and he proceeded from the assumption of the presence of a simple compressive process; however, this does not correspond to the physical essence of the phenomenon. As was explained in Chapter 8, in the case under review, ice pressure on a pier can be found from the equation

$$P = mF_{CM}R_{CM}.$$

As we shall see, pressure is directly proportional to the value m of the form coefficient, adopted on the basis of the findings of the plasticity theory and our experimental studies, fluctuating in the limits indicated in Table 36.

In this manner, applying a tapered face in place of the semicircular outline of cutting face, we can reduce the amount of pressure on a pier by 1.5 to 1.7 times.

We should review separately the question concerning the effect of a pier's form in plane in the presence of a sloped ice-cutting edge. An analysis of Eqs. (11.14) and (11.30) shows that with sloped ice aprons, a somewhat greater effect should be expected with blunter outlines of cutting face in plane.

Table 36

Values of Coefficient m

Angle of pier's tapering in degrees.	120	90	75	60	Semicircular
Coefficient m	0.81	0.73	0.69	0.65	0.90

In this case, the length of shearing area (or width of ice chunk being broken off during bending) decreases accordingly and hence less force is needed for disruption of the ice pack. The degree of influence of the pier's form on the value of effectual forces with sloped ice aprons can be explained as follows.

Let us utilize analogous formulas for finding the pressure of ice on structural supports during crushing, shearing and bending:

$$H_p = K_p b_0 h;$$

$$H_{cp} = K_{cp} b_0 h;$$

$$H_{u} = K_{u} b_0 h.$$

From these equations, it is easy to find the relationship:

$$r = \frac{H_{cp}}{H_p} \approx \frac{R_{cp}\, \text{tg}\, \beta \sqrt[3]{v}}{2 R_{cx} \sin \alpha \sqrt{\sin \alpha}} \qquad (13.1)$$

and

$$p = \frac{H_{u}}{H_p} = \frac{R_{u} S_0 \text{tg}\, 3 \sqrt[3]{v}}{1,6 R_{cx} \sqrt{\sin \alpha}}, \qquad (13.2)$$

by analyzing which we can determine the extent of decrease in force H at various tapering angles of the pier.

At the usual conditions of spring ice passages in the rivers of the North and Siberia (at $R_{compres}$ = 50 tons/m^2, $R_{и}$ = 50 tons/m^2, R_{cp} = 15 tons/m^2, v = 1 m/sec and b_o/h = 6), the values of coefficients r and p at various α-and β-angles are listed in Table 37.

Table 37

Values of Coefficients r and p

β $\diagdown$ 2α	During shearing(r)			During bending(p)		
	60°	90°	120°	60°	90°	120°
45°	0.43	0.25	0.10	0.15	0.12	0.10
60°	0.74	0.43	0.32	0.30	0.23	0.20
75°	1.60	0.93	0.69	0.55	0.72	1.24

The tabular data show what percentage is formed (by the horizontal shearing and bending pressure) of the pressure existing during the floe's crushing, but the latter itself depends on the pier's tapering angle. After conversion, the final data (Table 38) were obtained.

Table 38

Value	tapering angle in°			Remarks
	120	90	60	
Horizontal force during crushing..........................	100%	90%	80%	Based on Table 30
Horizontal force during shearing, in % of number of previous lines........................	0.19	0.25	0.43	At β = 45
The same, in % of pressure, at 2α = 120°	19	22	34	
Effect of pier's form in plane in % on pressure at 2α = 120° and β - 45°	100	116	179	
The same, at β = 60°	100	121	184	
The same, at β = 75°	100	122	185	

As we shall see, in the breakdown of an ice pack from shearing, most efficient are the piers with a minimal perimeter of shearing line. A similar calculation for the case of ice breakdown from bending has been presented in Table 39.

Table 39

Effect of Pier's Form During Bending

β \ $2\alpha°$	120	90	60
45	100	108	120
60	100	105	125
75	100	113	149

Thus, while for supports with a vertical face, sharper formations of the cutting face are desirable, with sloped ice aprons, they are somewhat less effective in case of ice breakdown from shearing. Such a finding agrees with the operating experience with wooden ice aprons of varying form.

G. Karlsen and N. Streletskiy [68], with the ice aprons of the hip type, observed the jamming of floes, which in 1944 led to the destruction of 4 ice aprons even under conditions of very light ice passage. The application of ice aprons of constant width with a smaller tapering angle proved more successful. A.A. Trufanov arrived at the same conclusion, and so did G.N. Petrov [67] on the basis of laboratory research.

In the breakdown of an ice cover by bending, the effect of the support's form in plane has little significance and is within the limits of accuracy of computing the ice loads (10-20%).

In this case, therefore, the form of a pier in plane should be selected, proceeding from the design concepts; special attention should be paid to the choice of the cutting face's inclination.

In choosing a pier's form in plane, one should not forget the conditions of flow around the piers by a stream or that the sharper piers split the ice more easily.

2. Case of Stopping Small Floes

The small floes, stopping at the piers, cut into the ice for a certain depth x_0, depending on the floe's kinetic energy reserve, ice strength and pier's form in plane.

As is known, the sharper piers are capable of cutting for a greater depth into the ice and hence the stopping of a small floe will occur with a smaller value of crushing area ω_{cM}.

Thus we are compelled to take the pier's form into account twice, as it were: in the replacement of the actual, complex process of the crushing phenomenon by the occurrence of compression by way of introducing the coefficient m of form and in determining the crushing area.

Specifically, for piers with triangular outline of cutting face, we have

$$P = k_0 vh \sqrt{\Omega\, R_{compres}\, m \tan \alpha}\ .$$

The force of ice pressure varies with variation in pier's form, just as the value.

$$z = \sqrt{m \tan \alpha}. \tag{13.3}$$

As is evident from Table 40, the force of ice pressure changes significantly for various forms of pier in plane.

Table 40

z-Values Based on Equation (13.3)

Tapering angle of pier in °	120	105	90	75	60	Semicircular
Value.........	1.8	0.99	0.84	0.71	0.59	1.57
The same, %	100	84	71	60	50	133

3. Case of Splitting the Floes

As the analysis conducted in Chapter 10 has shown, the effect of the tapering angle in this case also is quite significant in essence it establishes the nature of the stressed state.

From Eqs. (10.14), (10.15) and (10.17), it is evident that the force required for splitting a floe varies proportionally:

at sideways splitting of a floe....... to tan α /2,

at partial splitting .... to sin α, and

at breaking of the floe........... to coefficient n.

Table 41

Effect of Pier's Form During Splitting, %

Nature of disruption	Tapering angle in °				
	120	105	90	75	60
Partial splitting......	100	91	83	71	58
Splitting sideways.....	100	85	71	55	45
Breaking..............	100	84	71	58	47

In accordance with this, in Table 41 we have shown the effect of a pier's form in plane on the amount of interaction forces during a floe's splitting.

As we observe, even in the case of splitting, the effect of a pier's form in plane is quite appreciable.

4. Brief Conclusions

1. A pier's form in plane has a significant effect on the value of interaction forces developing between a pier and a floe.

2. In the presence of a vertical cutting face, it is desirable to use smaller angles of pier's tapering, which leads to a decrease in the forces of interaction between the pier and floe.

3. For a triangular outline of cutting face in plane (in %), the degree of reduction constitutes, for a tapering angle

equalling 120°, 90° and 60°, respectively:

at cutting through a large floe...... 100, 88, 75;

at stopping of floe 100, 71, 50;

at splitting of floe................ 100, 71, 47.

4. Thus, one should abandon the use of blunt configurations on a vertical ice-cutting face and transfer more boldly to the use of sharp outlines of piers, with the minimally possible tapering angle (based on design concepts).

5. In the piers with a sloped ice-cutting face, disrupting the ice sheet by way of bending, the effect of its form in plane is less significant and falls within the accuracy limits of estimating the value of the design loads.

6. In choosing the type of piers, one should not forget the conditions of their streamlining, in connection with which the solid piers of the major structures with sloped ice aprons should nevertheless be equipped with a tapered face. The form of the pier's tapered part should correspond to the water flow regime.

7. In the installation of wooden ice aprons (pile or cribwork type), it is advantageous to utilize more or less blunt outlines in plane, designed for breaking down the ice pack by shearing or bending. The use of starlings having constant width is desirable; this eliminates the possibility of the floes' jamming between them.

8. The piers of semicircular configuration can find application with a sloped ice-cutting face and should not be used in piers with a vertical cutting face.

FOURTH PART

PRESSURE OF ICE ON STRUCTURES OF CONSIDERABLE EXTENT,
AND OTHER CASES

Chapter XIV

STATIC PRESSURE OF ICE ON STRUCTURES

1. Features of the Phenomenon

Static pressure of ice on structures develops chiefly at
a rise in temperature of the ice pack, when its free deformations
obstruct the reaction of shores and structures. This type of
action acquires the maximum significance for the structures of
considerable extent (dams, dock facilities, cofferdams etc.)
under climatic conditions typified by abrupt fluctuation in air
temperatures and hence in temperatures of the ice pack.

The analysis of the elastic-plastic properties of the ice
pack presented in Chapter 5 permit us to note that at the present
level of knowledge, the theoretical developments of the question
are extremely difficult. The plastic deformations of ice are
in complex dependence on a number of factors, including temperature and structure of ice, duration of effect and nature of
loads, and form of stressed state. In addition, a tremendous
influence, quite difficult to take into account, is exerted by
local conditions, specifically by the configuration and nature
of the shore line.

Therefore, at the present time, there are still no established methods for estimating the value of static pressure of
an ice pack, and natural observations coupled with carefully
formulated experiments still have special significance.

2. Temperature Regime of Ice Sheet

Under normal conditions, the temperature regime of the
ice pack is very complex. If the lower surface of the ice has
a temperature of 0°C, temperature of the upper surface depends
to some extent on the variation in air temperatures, and on

thickness and density of the available snow deposits. The temperature regime of the central part of an ice pack is determined by its thickness, the thermal-physical characteristics of the ice and also by the pattern of air temperatures.

Of definite interest are the extensive observations carefully undertaken by G.Ya. Kuzub [149] on the temperature regime of the ice pack of one of the Siberian rivers, conducted by him during two winters. As might have been expected, the diagrams of the temperatures' distribution can be represented by three types:

1. Temperature of ice surface below the temperature of the subjacent layers. This type occurs during cooling and is especially common at the beginning of the winter period. The average ice temperature in respect to depth is usually found at a distance of 0.45 h from the surface of the ice pack.

2. Temperature of the ice surface is higher than temperature of subjacent layers. This type occurs in the period of a rise in air temperatures and is customary for the spring period. The average ice temperature by depth is located at 0.56 h from the surface of the ice pack.

3. The third type is a transitional one between the first and second and can occur during the entire winter. Let us note that the linear distribution in temperatures occurs quite rarely.

An analysis of the relationship curves of mean diurnal air temperatures and mean temperature of ice $t_{\varPi}$ permitted us to conclude that the latter is established mainly by air temperature t_{β}, taken for the preceding days and can be found with the empirical relationship:

$$t_{\varPi} = 0.32t_{\beta} - 1°, \qquad (14.1)$$

from which it follows that the ratio of ice and air temperatures is most constant but varies as follows:

t_{β} for previous days...	-25°	-10°	-15°	-20°	-30°C
$t_{\varPi}:t_{\beta}$	0.52	0.42	0.38	0.37	0.35

Of interest are the data concerning the temperature fluctuations which have occurred in nature in the depth of the ice pack during the days and their comparison with the fluctuations in air temperature for the preceding days (Table 42).

Table 42

Temperature Fluctuations in Layer of Ice Pack

Data	t_3	Distance from surface of ice pack, cm					
		1-3	17	35	50	65	80
Amplitude of fluctuations for day in °C.................	7.5	6.5	3.1	1.9	1.6	1.2	0.4
The same, %.....................	100	87	42	25	21	16	6
No. of observations............	119	158	105	110	100	84	64

Thus, the most intensive attenuation of temperatures occurs in the upper "active" layer. Fifty percent of the amplitude is attenuated in the upper layer with a thickness of 10 cm, while only 25% of the amplitude reaches a depth of 35 cm.

The snow pack exerts an ameliorating effect on the temperature fluctuations of the ice. It is noted that in the presence of snow, the pattern of ice temperatures will lag behind the pattern of air temperature no longer by one, but by about 2 days.

In this case, the ice temperature can be determined with the equations suggested by G. Ya. Kazub [149]

$$t_\pi = 0,50 \frac{h_\pi t_\text{в}}{h_\pi + 9h_\text{c}}, \tag{14.2}$$

$$t_\pi = 0,49 t_\text{c} \frac{h_\pi}{h_\pi + 9h_\pi} - 1°, \tag{14.3}$$

where h_c, h_π = thickness of snow and ice; $t_\text{в}$ = average of the mean diurnal air temperatures, °C, taken from the previous 2 days.

An analysis of the distribution in temperatures in the column of an ice pack and their fluctuations in time is given in reports by B.V. Proskuryakov [64] and N.N. Petrunichev [62,104].

The results of these interesting studies reduce to the following:

1. In connection with the great complexity of the process and the diversity of the actual conditions, the initial distribution in temperatures is assumed for the steady thermal conditions.

2. Problems are formulated and solved, determining the
further variation in distribution in ice temperature through
time at various conditions of a rise in air temperature (jump-wise
or gradual, in presence or absence of snow on the ice).

3. Mathematical formulas and tables are given, permitting
a computation of the temporal variation in ice temperature. In
this connection, the thickness h_c of the snow deposits on the
ice is replaced by an equivalent layer of ice.

3. Survey of Methods for Establishing the Pressure

The experimental studies permitted Royen [63] to suggest a
relationship between relative ice compression ε , in mean temperature t_o, amount of load P_o and the time τ of its action in the
known form:

$$\varepsilon = \frac{C P_0 \sqrt[3]{\tau}}{t_0 + 1}.$$ (14.4)

In spite of the fact that (14.4), Royen proceeds from an
incorrect concept concerning the damping of deformations ε with
time τ, this formula has served as a basis for a widely accepted
method for determining the static pressure of ice during thermal
expansion. It can be established that the maximum stress

$$\sigma_{max} = 1640 \alpha (t_0 + 1) \sqrt[3]{\frac{t_0}{s} (t_0 + 1)^2} \ \kappa g/c m^2,$$ (14.5)

then, having assumed the coefficient of linear expansion of ice
$\alpha = 0.000055$, we can determine the ice pressure per running meter
based on the formula

$$P = 0.9 h (t_0 + 1) \sqrt[3]{\frac{t_0}{s} (t_0 + 1)^2} \ \frac{\tau/\tau. \, m}{m/nog. \, m.}$$ (14.6)

In Eqs. (14.5) and (14.6), t_o = average initial ice temperature in depth; s = time needed for raising the temperature from
t_o to 0°C; and t_o/s = the gradient of increase in ice temperature
through time (for a prolonged period of rise, s = 4-15 days).

Based on Royen's formula, an equation was derived which
superficially was quite similar to (14.6) but differed from it in
its nature. It is listed in GOST 3440=46, where t_o is no longer
the ice temperature but its maximum possible increase in the time
s.

As was indicated by B.V. Proskuryakov [64], this did not follow from the solution to the Royen equation and in addition, pressure became independent of the physical properties of ice, varying perceptibly with temperature, which is quite incorrect; and also in solving the Royen equation, at a differentiation in Eq. (14.4), a constancy was assumed in temperature and pressure, but this does not correspond to reality.

B. V. Proskuryakov suggested considering in a first approximation that ice behaves, in the area of plastic deformations, as a viscous liquid, and he developed a method permitting a determination of the calculated movement in dependence on a number of factors including the coefficient of ice viscosity (η). The utilization of this technique is made difficult by the fact that, as was explained later on [94,120,124], the coefficient of ice viscosity is not a specific physical constant but changes depending on numerous factors.

A number of refinements have been suggested by N.N. Petrunichev [104], however, as was correctly indicated by him personally "the absence of validated relationships between the deformation of ice, effective stress, time and temperature complicates the development of a reliable theoretical system for determining the maximal pressure of ice pack at prescribed meteorological conditions".

4. Field Determinations of Static Ice Pressure

The complexity of the problem greatly complicates the development of theoretical procedures and increases the importance of field observations.

From 1954-1956, the AUSRITH (All-Union Scientific Research Institute of Hydroengineering) [104] organized the measurements of stresses in horizontal beams at the Dneprovskaya HES imeni V. I. Lenin. The calculations permitted an estimation of the amount of pressure at 24.6 tons/m^2 (12.3 tons per running meter).

Of considerable interest are the studied conducted in the U.S. on five reservoirs in the mountains in Colorado [123,104]. For the determination, use was made of electric strain gauges and sensors, having functioned according to the Brinell principle, having been installed at three points through the ice layer.

The maximal pressure was obtained at the Eleven Mile Canyon Reservoir from 1947-48, 1948,49 and 1949-50 amounting to 39, 41 and 58 tons/m^2 (or 24, 21 and 30 tons/running meter) respectively, with steep, rocky shores.

In winter of 1950-51, determinations at four other reservoirs yielded 5, 9, 14 and 25 tons/running meter, respectively. The difference is explained by the outline and structure of the shores. Data are also available on damage to the sluice gate at the Keokuk Dam [165] with a pressure of 12 tons/running meter (21.5 tons/m^2), and concerning the wrecking of a dam at Minneapolis [60,43] with a pressure of 16.7 tons/m^2.

In this manner, the value of static ice pressure has an order of magnitude of 5-30 tons/m^2 in the case of shores permitting deformation and up to 60 tons/m^2 in case of steep, rocky shores.

5. Recommendations of SN 76-59

In a review of GOST 3440-46, the staff at the Laboratory of Ice Thermics at the AUSRIH under the supervision of A.M. Yestifeyev [193] on the basis of experiments and calculations suggested an equation of the form

$$P = A (t_{н.л} + 1)^{4/5} \theta^{1/5},$$

$$(14.7)$$

where $t_{н.л}$ = initial ice temperature; θ = gradient of temperature variation through time; and A = variable coefficient depending on $t_{н.л}$.

In the utilization of the actual studies and literature data [123], we selected the relationship of coefficient A with air temperature t_β :

$$A = 0.78 t_\beta^{-0.88},$$

$$(14.8)$$

providing the maximum P-values and corresponding to the gradients frequently occurring in various regions of the USSR.

As a result, we have recommended the formula:

$$P = 3{,}1 \frac{(t_{н.л} + 1)^{1.67}}{t_{н.л}^{0.88}} \theta^{0.33},$$

$$(14.9)$$

having entered SN 76-59. At practical application of the method, we assume

$$t_{\text{к.л}} = 0{,}35 t_{\text{в}} \ °C,$$

$$0 = \frac{0{,}35 \Delta t_{\text{в}}}{s} ,$$

where $\Delta t_{\text{в}}$ = the rise in air temperature, degrees, during time s expressed in hours.

In the presence of snow and at a considerable length L of ice pack, we introduce the correction coefficients (respectively):

$$r = \frac{h_{\text{л}}}{h_{\text{л}} + 9{,}1 h_{\text{с}}} , \tag{14.10}$$

$$\psi = 0{,}6 - 0{,}9 \, (\text{at} \ \ L = 150 - 50 \ \text{м}).$$

In SN 76-59, use is made of the Royen equation, incorporating certain refinements.

Based on the recommendations of SN 76-59 for 10 points in the country, the pressure value will fluctuate from 15 to 30 tons/m^2.

Undoubtedly, we require an additional development of the question and primarily an undertaking of natural (field) observations.

6. Static Pressure of Ice on Individual Buttresses (Piers)

The static horizontal pressure of ice on individual piers caused by its thermal expansion has significance only in the case of unilateral pressure. Here the question arises whether one should determine the pressure from an ice field by the width equalling half the sum of adjacent spans or be limited to consideration of only the pier width, as is suggested in reports [1, 166, 167].

However, as was already indicated by the author [54] and then by N.N. Petrunichev [62], such a procedure leads to a decrease in the value of calculated pressure. The deformation of the part of ice field lying between the piers (Fig. 49) does not occur separately and exerts a partial effect on the amount of pressure on the support, increasing it. Petrunichev [62] provided an approximate solution of this problem, proceeding from the system of an ice pack's behavior during thermal expansion, adopted by B.V. Proskuryakov.

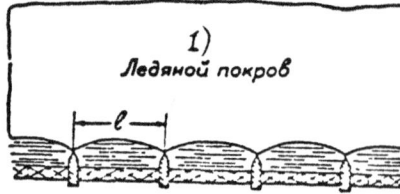

Fig. 49. On a Determination of Static Pressure of Ice on Piers. Key: 1) ice pack.

Assuming that all the stresses from the span could be transmitted to the ice located opposite the pier, and having made a number of other assumptions, Petrunichev derives the pressure on a pier

$$P_6 = P_{0_{max}} \left(b_0 + \frac{l}{3} \right),$$

(14.11)

where $P_{0_{max}}$ = maximal pressure per running meter of structure in the case of solid resting of ice; b_0 = width of pier; and l = the half-sum of the spans contiguous with the pier.

Obviously, as a first approximation, we should agree with Petrunichev's recommendations, at this time having indicated that the pressure can not be greater than that critically possible (proceeding from the crushing or plastic flow of ice), i.e. 35-40 tons/m^2.

7. Brief Conclusions

1. Static pressure of ice on structures is closely related to the still insufficiently studied plastic properties of ice and the actual conditions (nature and structure of shorelines, thermal regime of ice, meteorological conditions, presence of snow on the ice, and so forth).

2. In connection with this, a generally recognized calculation technique has not yet been developed.

3. The experimental observations indicate the moderate value of the developing forces of interaction (15-40 tons/m^2), in connection with which for the trial calculations, we can utilize the recommendations in SN 76-59.

4. Unilateral pressure on separately standing piers can be estimated according to the suggestion made by N.N. Petrunichev.

5. It would be extremely desirable to make further studies and particularly to undertake observations under natural conditions

Chapter XV

DYNAMIC PRESSURE OF ICE ON STRUCTURES OF GREAT EXTENT

1. General Remarks

During the period of spring ice debacle, it is possible to have the interaction of large ice fields (having been brought into motion under the effect of current or wind) with structures of considerable length, i.e. dams, cofferdams, dock structures, embankments or the slopes of river banks.

In this connection, an ice field having a known reserve of kinetic energy expends it chiefly in breaking up its edge, at this time exerting a certain amount of pressure on the structure.

At contact with a vertical (or nearly vertical) face of the structure, the ice pack breaks up from crushing. At a considerable inclination of the pressure face of the structure to the horizon ($\beta < 75°-80°$), it is possible to have breaking from bending. After the approach of a large ice field, one observes its partial disruption, accompanied by climbing up or the accumulation of floes at the structure, and the subsequent action of ice is already absorbed by the ice bank which has formed. In connection with this, the breakdown of floes from shearing (usually occurring in the later phases of ice passage) can be overlooked.

Often an ice field can move at angle φ to the pressure front of the structures, which also should find reflection in the calculation techniques.

2. Effect of Ice on Structures With Vertical Face

This question has been reviewed by a number of the authors [1, 43, 47, 51, 52, 59, 76] and has received definite representation in GOST 3440-46 and SN 76-59.

The suggestions made by N.M. Shchapov [41], then developed by P.A. Kuznetsov [47], by the author [53,54] and B.V. Zylev [163] appear to be those which correspond most to the physical pattern of the occurrence.

Recommendations of SN 76-59

The technical specification for finding the ice loads on river structures (SN 76-59) recommend in this case the use of the formula

$$H = kvh\sqrt{\Omega},\qquad(15.1)$$

when at

$$R_{compres} = 30 - 50 \text{ tons/m}^2 \qquad k = 2.36 - 3;$$

$$R_{compres} = 60 - 100 \text{ tons/m}^2 \qquad k = 3.3 - 4.3.$$

The maximal H-value in the plane of contact with the structure should be assumed proceeding from the ultimate compressive strength of ice equalling 30-100 tons/m² depending on the stage of ice passage and the climatic features of the region.

Suggestion Made by the Author

A floe of rounded or roughly rectangular outline with dimensions Ω, h will move toward a structure with velocity v, having a known supply of kinetic energy. At contact of the floe with the structure, most likely we will have the impact of one of the edge projections, pointed or rounded in plane.

It is obvious that the action will be quite analogous to that reviewed in Chapter 9 but here we can get by without considering the local crushing or the coefficient m of pier's form. Therefore the working formulas acquire the form:

a) for a floe, the impact of which is reflected in Fig.35 based on Eqs. [53,58]

$$P = 0.31\ vh\sqrt{\Omega R_{compres}(\tan\alpha + \tan\beta)};$$

b) for a floe, the impact of which is reflected in Fig. 35, b,

$$P = 0.43\ vh\sqrt{\Omega R_{compres}\ \tan\alpha};\qquad(15.2)$$

c) for the floes of rounded outline (at $2\alpha = 140°$)

$$P = 0.71\ vh\sqrt{\Omega R_{compres}}.\qquad(15.3)$$

From a comparison of these equations, it is evident that Eq. (15.3), yielding the maximum pressure value for the frequently occurring floes of rounded shape should be the design (calculation) equation.

Table 43

Recommended Values for $R_{compres}$ k

Velocity,m/sec	$R_{compres}$,t/m^2	k	$R_{compres}$k
0.25	80	0.6	50
0.50	65	0.7	45
1.00	50	0.8	40
1.50	40	0.9	35

Let us note that the critical pressure of length cannot be greater than

$$P_{max} = R_{compres}kh, \qquad (15.4)$$

where k is the coefficient of nondensity osculation.

As was indicated in Chapter 9, the $R_{compres}$ and k-values are linked with the velocity of floe (Table 43).

In this manner, the marginally possible ice pressure cannot exceed:

$$P_{max} = (35-50) \; h \quad t/m^2.$$

Comparing our recommendations with the requirements imposed by SN 76-59, we note that in the Technical Specifications, Eq. (15.2) at 2 α = 90° is adopted for a basis; the great pressure from rounded floes is not considered and the ultimate crushing limit of ice is exaggerated somewhat.

3. Effect of Ice on Structures with Inclined Face

The horizontal component based on SN 76-59 for inclined planes is found with the equation

$$H = q \tan \beta = R_{M} \lambda' h^2 \tan \qquad (15.5)$$

at $R_{\text{и}} = 105-31$ tons/m^2 depending on the ice passage conditions and the λ'-values associated with the ice thickness (Table 44).

Table 44

Values of λ'

h, m	0.4	0.5	0.6-0.7	0.8-0.9	1.0-1.3
λ'	0.08	0.07	0.06	0.05	0.04

The vertical pressure component is assumed to equal

$$q = R \lambda' h^2 \tag{15.6}$$

Equations (15.5) and (15.6) have been developed on the basis of the studies by B.V. Zylev [45], analyzed in detail in Chapter 11. Let us recall that Zylev, examining the question concerning the pressure of ice on sloped ice aprons, failed to consider their form in plane, which does not have significance for the given instance. Let us examine the question in more detail.

As is known, in this case after the contact of an ice field with a structure, we observe the crushing of the lower floe's edge (see Fig. 42).

At inclination of the structural face to the horizon at the angle $\beta < 75°$, the breaking of the ice field edge develops for a distance of about 3 h from the contact zone. The fluctuations in the strength of ice pack and the roughness of the structure's edge lead to an additional splitting of the ice pack into individual floes (often of a square or rectangular form in plane) which, under the pressure of the moving ice, climb onto the slope and accumulate.

Applying in a first approximation the usual equations from structural mechanics

$$R = M/W \, ,$$

we find (at $M = P \cdot 3h$ and $W = bh^2/6$) the vertical force

$$V = R_{\text{и}} \ h/18$$

and corresponding to this, the horizontal pressure

$$H = 1.IV \tan \beta = 0.062 R_и h \tan \beta \quad , \tag{15.7}$$

Taking into account the possible local increase in the thickness and strength of ice, we suggest that it is feasible to assume finally:

$$H = 0.1R_и h \tan \beta \tag{15.8}$$

at $R_и = 40-65$ t/m^2.

We have listed in Table 45 the results of calculations using Eq. (15.8) according to the requirements of SN 76-59.

Table 45

Comparison of Calculations Based on (15.8) and (15.5)

| Method | $R_и$, t/m^2 | Angle β , in degree, of inclination to horizon | | | | Remarks |
		30	45	60	75	
According to SN 76-59	105	2.4 1.1	4.2 1.8	7.3 3.2	15.0 6.9	h = 1 m h = 0.5 m
According to Korzhavin	65	3.8 1.9	6.5 3.3	11.3 5.7	24.3 12.2	h = 1 m h = 0.5 m

Based on our suggestions, we derived somewhat higher values for the calculated ice pressures, which condition apparently is closer to reality.

A.I. Gamayunov [156] considered the stressed state of an ice field as in a semi-infinite isotropic plate lying on an elastic base. He proposed the following equation for finding the horizontal force per running meter (r.m) of the structure's length:

$$H = 0.52R_и m \lambda h^2 \tag{15.9}$$

where $m = \dfrac{1 + f \cot \beta}{\cot \beta - f}$ = the coefficient taking into account the friction and inclination of structure's face to the horizon;

f = 0.11 = coefficient of ice friction on concrete;

$$\lambda = \sqrt[4]{\frac{r}{4D}} = 0.169 - 0.074 \text{ (for } h = 0.40\text{-}1.20 \text{ m)};$$

r = pliability factor of elastic base;

$$D = \frac{Eh^3}{12(1-\mu^2)} = \text{cylindrical stiffness of plate;}$$

μ = Poisson coefficient; and

E = elastic modulus assumed to equal a total of 5 tons/m^2.

Analyzing Gamayunov's data, we find that the λ-values recommended by him will fluctuate in the limits ranging from 1/14.8 h to 1/11.3 h. In connection with this, the breakdown begins (even at the assumed E-values) at distance x from the plane of contact with the floe, equalling:

$$x = \pi/4\lambda = (8.8 \div (11.6) \text{ h}.$$

With the values

$$x = 10 \text{ h}, \lambda = \pi/4x = 0.0785h^{-1}, f = 0$$

expression (15.7) assumes the form:

$$H = 0.041R_{\text{и}} h \tan \beta , \qquad\qquad (15.10)$$

which fully corresponds to the structure of Eq. (15.8).

The calculations based on Gamayunov's Eq. (15.9) provide the values of calculated pressure expressed in tons per running meter of structure (at h = 1 m) for the various β-values (Table 46).

Table 46

H-Values, After A.I. Gamayunov

steepness of slope	1 : 1	1 : 1.5	1 : 3	Remarks
	45°	32°40'	18°20'	
P_{max}, t/r.m.	4.0	2.4	1.5	$R_{\text{и}}$= 75 t/m^2
P_{min}, t/r.m.	2.2	1.3	0.8	$R_{\text{и}}$= 40 t/m^2

Considering further that a) the phenomenon in effect passes beyond the limits of elasticity, b) the assumed E-value is far from actuality, c) in nature, the breaking occurs at a considerably shorter distance than 9 - 11 h, d) the consideration for the frictional forces is situated within the accuracy limits (10-20%) of the method and e) the values derived for P_{max} are depressed, we consider it possible to recommend our Eq. (15.8) for application; it is simple in structure and is based on the studies of the physical pattern of the occurrence in nature. In addition, the calculations based on this formula provide somewhat higher values of the ice loads, which for the structures in the North and Siberia is a quite valid situation.

4. Movement of Ice at Angle to Front of Structure

During the drift of an ice field at angle φ to the front of a structure, the phenomenon becomes complicated. In a first approximation (assuming that the entire reserve of kinetic flow energy is spent only in the work of demolishing the ice), we have: a force perpendicular to the front of the structure (Fig. 50).

$$P_1 = P \sin \varphi ; \tag{15.11}$$

a force parallel to the front of structure,

$$P_2 = P f \sin \varphi , \tag{15.12}$$

where P = pressure on structure, perpendicular to the direction of movement; φ = angle between direction of floe motion and front of structure; and f = the coefficient of friction between the floe and structure equalling an average of 0.11 (0.07-0.14). In SN 76-59, we suggest also taking into account the energy expended in turning the floe during the impact process.

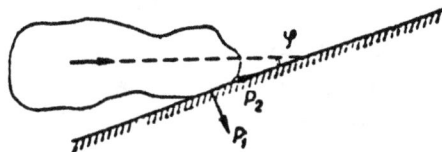

Fig. 50. Case of Action of Floe Drifting Obliquely to the Front of a Structure.

At this time, Eq. (15.11) becomes greatly complicated and acquires the form:

$$P_1 = c v h^2 \sqrt{\frac{\Omega}{\mu \Omega + \lambda h^3}} \sin \varphi. \tag{15.13}$$

In our opinion, one should have replaced it by a simpler formula such as

$$P_1 = aP \sin \varphi$$

and provide a nomogram of the formula

$$a = \frac{c}{k \sqrt{\lambda + \frac{\mu\Omega}{h^3}}}$$

in relation to φ, Ω, h and $R_{\text{и}}$.

It also appears that in estimating the order of magnitude of ice pressure, it is possible to disregard the work of the forces causing the rotation of ice and assume a = 1.

5. Brief Conclusions

1. During the driving of a large ice field by wind or current onto a sizeable structure, the value of horizontal P or vertical V pressure components should be found with the equations

a) with vertical face

$$P = 0{,}71 vh\sqrt{\Omega R_{\text{сж}}} \leqslant P_{\max},$$

$$P_{\max} = R_{\text{сж}} hk = (35 \div 50) \, h \quad t/r.m.$$

b) in case of sloped face:

$$P = 0.1 R_{\text{и}} h \tan \beta \ ,$$

$$V = P \cot \beta \text{ at } R_{\text{и}} = R_{\text{compres}} = 40\text{-}65 \ t/m^2.$$

2. The requirements of SN 76-59 have been developed for a particular case (impact an an angle equalling 90°, static nature of force's application) and elevated limits of ice strength.

3. At movement of floe at angle φ to the front of structure, one should assume the force $P_1 = P \sin \varphi$ and $P_2 f \varphi P \sin$ at coefficient of friction on concrete $f = 0.11 \ (0.7^2 - 0.14)$.

Chapter XVI

OTHER CASES OF ICE ACTION

1. Action of Ice Sheet Frozen to Structure

During ice passage, we find the freezing of ice with the structural elements, both with those which are extended (slopes of dams) as well as those which stand separately (piles, sumps etc.). At fluctuations in water level, it is possible to transmit through the frozen ice sheet to the structure the vertical forces acting upward or downward.

A. Case of Separately Standing Structures

At an increase in the water level, it is possible to have the transfer of hydrostatic pressure through the frozen ice pack to a separately standing support. A.N. Komarovskiy [1], considering a precise solution difficult, suggested estimating the value of the developing force, proceeding from the possibility of shearing around the structure's perimeter (at ultimate strength of 90 tons/m^2). Taking into account that the breakdowns of an ice pack are also possible at a certain distance from the structure, Komarovskiy also suggested determining the calculation force, proceeding from the operating conditions of ice for bending at a strength limit of 300 tons/m^2.

P.A. Kuznetsov [47] considered it possible to visualize an ice pack as a round plate, loaded from beneath by a uniformly distributed load. In Gamayunov's [158] opinion, Kuznetsov assumed (without enough justification) the diameter of the calculated plate to be 50 h.

As a result, Kuznetsov suggested the relationship

$$P_s = \frac{300\,h^2}{\ln \frac{50h}{d}},$$ (16.1)

having been incorporated in GOST 3440-46 and later also in SN 76-59, where h = ice thickness, m; and d = diameter of pile or section, m.

With a rectangular form of section with sides a and b, the d-value is assumed to equal $\sqrt{ab}$.

A.I. Gamayunov [158] provided a more complete solution to the problem, having reviewed the bending of a thin round plate with external radius R, finished elastically with respect to the outer and inner outlines (Fig. 51).

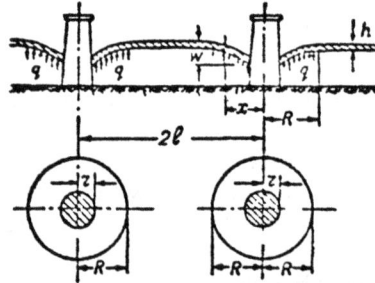

Fig. 51.   Pattern of Action in Case of a Frozen Ice
Sheet, Assumed by A. I. Gamayunov.

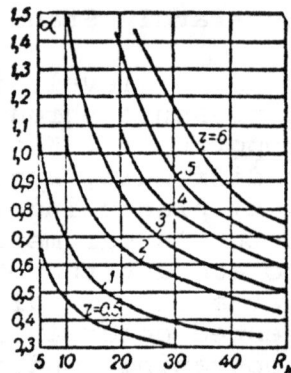

Fig. 52.   Curves for Finding the Coefficient α in the
Gamayunov Equation.

The load which is acting upward is assumed to be evenly distributed.

As a result of a detailed analysis [158], we have derived the working formula of the type

$$R_B = \alpha R_и h^2 \tag{16.2}$$

where

$$\alpha = \frac{\pi (R^2 - r^2)}{6A}$$

-202-

and

$$A = \frac{R^4\left(4\ln\frac{R}{4}-3\right)+r^2\left(4R^2-r^2\right)}{8\left(R^2-r^4\right)} \ .$$

For utilizing Eq. (16.2), one must know the radius of load distribution R which we recommend finding with the equation

$$R = 3{,}92\sqrt{\frac{Er^3}{12(1-\mu^2)}}. \qquad (16.3)$$

The calculations of the R-value for various ice thicknesses yield the following results (at E = 30 t/cm^2):

h = 0.25	0.50	0.75	1.00	1.25	1.50 m
R = 15	30	40	51	60	70 m

It is easy to note that the R-value roughly comprises 46-60 ice thicknesses. For calculating the α-value, Gamayunov recommended the graph presented in Fig. 52.

A comparison of the suggestions contained in SN 76-59 and made by Gamayunov is made below for the following initial data:

d = 2r = 2 m; E = 30 t/cm^2; h = 1 m; t = -6°C.

The calculated value of strength limit for the temperatures differing from zero can be found with the equation

R = 11 + 3.5 t = 32 Kg/cm^2 = 320 t/m^2.

Comparing the calculations based on Eqs. (16.1) and (16.2), we have:

after (16.1) $\quad P_s = \frac{300\,h^2}{\ln\frac{50h}{d}} = 93$ tons.

after (16.1) $\quad P_B = R\,\alpha h^2 = 0.33 \cdot 320 \cdot 1 - 105$ m

i.e., in effect, quite similar results.

B. Ice Actions on the Concrete Coverings of Slopes

In the presence of fast shore ice (pripay) and fluctuations in water levels, we find cases of the disruption in the stability and strength of the concrete revetments along dam slopes.

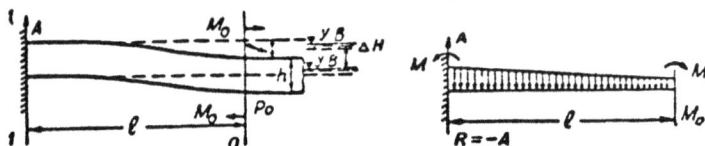

Fig. 53. Pattern of Deformation of Frozen Ice Pack and Its Effect on Concrete Revetment of Slope (after P. A. Shankin).

The physical pattern of the occurrence reduces to the following. At a drop in water level, the frozen ice pack begins to be deformed, causing the appearance of bending moment M and of vertical force R according to the system indicated in Fig.53. At this time, we have plastic deformations of the ice pack and the appearance of cracks running parallel to the slope.

P. A. Shankin [159,160] suggested a method for solving the complex problem, to be sure based on appreciable assumptions. Considering ice as an isotropic elastic body, Shankin wrote a differential equation of the cambered axis of an ice pack and after its integration, he found an expression for the reaction of support R, bending moment M, and he determined the position of a section with maximal sagging, l.

After several simplifications for a specific case (rate in decrease of water level 1-3 cm/hr, $\psi = 1.0$, we can derive: bending moment in section of "closing" (1-1)

$$M = R_{\text{и}} h^2 c/6; \tag{16.4}$$

the vertical reaction in section of closing

$$R = \frac{1,37\,M}{\sqrt[4]{\frac{EI}{c\gamma}}}; \tag{16.5}$$

the distance between sections 0-0 and 1-1

$$l = 3,58 \sqrt[4]{\frac{EI}{c\gamma}}; \tag{16.6}$$

the moment in section 0-0

$$M_0 = \frac{2cl^4\gamma - 360EI}{3cl^4\gamma + 360EI} M,$$

(16.7)

where h = ice thickness; E = elastic modulus of ice; c = width of ice pack section; I = inertial moment of ice pack section; and γ = volumetric weight of water.

According to Shankin's data, the equations obtained confirm satisfactorily the natural observations undertaken at the Yakhromskoye Reservoir at E = 40 t/cm^2.

Evaluating the proposals made by Shankin, we should note that in connection with the many assumptions (the ice is assumed to be an elastic, isotropic body, the allowance for the effective time of load is conducted roughly, and the phenomenon of ice relaxation was not taken into account), they can be regarded as a very initial approximation and undoubtedly require refinement.

Natural(field) observations are also extremely desirable.

2. Pressure Developing During Accumulations of Ice at Structures

In the case of severe ice passages and also in the process of the formation of jams and buildups of water under snow, the ice accumulates at the shores and structures. Cases are on record of the accumulation of ice to a height of 30 m [1], while some pileups of 10-15 m are a fairly frequent occurrence along the large rivers in Siberia [76].

In addition to the shore accumulations, it is possible to find piles of ice at the crest of overflow dams with a slight thickness of overflowing layer or improper shaping of crest. In the period of spring ice passage at the Volkhovskaya Dam, accumulations with a height up to 4.5 m [161] are observed.

It is obvious that in this case the ice pressure can be computed with the formula

$$P_\beta = 0.7H \text{ t/m}^2$$

(16.8)

where P_β = the vertical pressure component, t/m^2; H = height of pile, m; and 0.7 = volumetric weight of accumulation which, according to the suggestion by V.M. Samochkin [76] can be assumed to equal 0.7 t/m^3.

In case of the pileup of ice on the sloped faces, it is also possible to have the appearance of a component P_H (Fig. 54) perpendicular to the slope.

$$P_H = P_\beta \sin \beta , \qquad (16.9)$$

where β = angle of face's inclination to the horizon. Let us point out that in this case, it is also possible to have the appearance of additional horizontal forces from wind action on the ice accumulations.

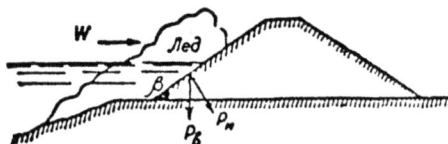

Fig. 54. Diagram of Forces Developing During Accumulations of Ice.   Key: 1) ice.

As we observed, in the case under review, the ice pressure is slight and can be of interest only for designing the light structures of the trestle type bridge.

3.   Abrasive Effect of Ice

If the ice fields during the ice passage move along the front of a structure (embankment, cofferdam, dam), coming in close contact with them, damages are possible, including the appearance of furrows of some given depth at the level of the ice passage.

We are aware of cases of the destruction of shore facilities [82] from the appearance of deep furrows (1.5 X 0.3 m) cutting through piles [162] in 5-6 hours and damaging the shore revetments. During the ice passage in 1960, at the Mamakanskaya HES [148], the abrasion of a layer of concrete for 3-7 cm was recorded. During uniquely severe ice passage, similar grooves are formed even on the rocky cliffs of certain Siberian rivers.

It is obvious that the calculated value of longitudinal ice pressure can be determined as follows.

The maximally possible ice pressure perpendicular to the front of a structure (with a vertical wall) can be found with the formula

$$P_H = R_{compres} hk \qquad (15.4)$$

or P_H = (25-50) h t/running meter. Obviously, the longitudinal ice pressure can be derived based on the equation

$$P_{np} = P_H f \qquad (16.10)$$

at frictional coefficient f = 0.11 (for the friction of ice on concrete).

Hence, the value of longitudinal pressure

$$P_{np} = (2.7-5.5) \text{ h t/running meter} \qquad (16.11)$$

while for the rivers of the European sector of the USSR, it can be reduced by as much as half. Let us note that Komarovskiy [1] estimated the value for the longitudinal pressure for the Svir' R. at 0.5-0.6 t/running m, i.e. with a value of the same order of magnitude.

The calculations made by Samochkin [76] for the Yenisey R. led to the choice of a calculation pressure of the order of 2.1 t/running m.

It is necessary to have in mind that even the low pressures at prolonged action can cause the abrasion of material. In connection with this, we could recommend the development of a smooth surface for the walls of the structure, an increase in the strength of surface layer of concrete, the application of facing and the use of sloped walls. The last measure can reduce significantly the design thickness of a floe and cause the accumulation of ice, which will reliably protect the structure from further damages.

4. Pressure of Freely Floating Ice Field During Pileup

After the approach of an ice field to a structure, one observes the breaking of its edge and then the stoppage of the ice floe. However, even after stopping, the field is capable of exerting pressure owing to the action of current and wind on the rough surfaces of the ice pack.

These questions have been studied to some extend by B.V. Proskuryakov [172], N.N. Petrunichev and I.M. Mamayev [173], A.S. Ofitserov [174], by the author [53], P.A. Kuznetsov [47], G.V. Polukarov [175] ,A.M. Latyshenikov [176],N.N. Zubov [177], A.I. Pekhovich, S.M. Aleynikov, N.F. Buznikov and others.

As a result of a review of GOST 3440-46 by the staff at the Ice Thermics Laboratory of the AUSRIHE, a formula was proposed which was then incorporated in SN 76-59. This formula (for finding the horizontal pressure perpendicular to the structural front P_H) has the following form:

$$P_H = \Omega[(\rho_1 + p_2 + p_3)\sin\varphi + p_4\sin\beta]\ Kg, \qquad (16.12)$$

where $p_1 = 0.5\ v^2$ = the force caused by flow friction on the lower surface of ice, Kg/m^2; $p_2 = 50hv^2/L$ = force of pressure, Kg/m^2, on ice pack's edge, referred to unit of floe's area; $p_3 = 920\ hi$ = horizontal component of force caused by influence of slope (i); $p_4 = (0.001 - 0.002)$ = the force caused by the wind-caused flow w in m/sec on surface of ice pack; φ = angle between front of structure and current direction; β = angle between front of structure and wind direction; Ω, h, L = dimensions of floe.

It seems to us that Eq. (16.12) is excessively complicated. In actuality:

1. Wind direction is variable and it is always possible to assume the most disadvantageous value of β (equalling 90°).

2. The effect of hydrodynamic pressure (on the mid-section of a floe with area Bh) is relatively slight and can be overlooked. The error developing is within the accuracy limits of the calculation. In connection with this, we can assign a simpler form to Eq. (16.12):

$$P_H = \Omega[(p_1 + p_2)\sin\varphi + p_4]\ Kg. \qquad (16.13)$$

For the wind speed value, we should recommend definite limits, since the maximum speed assumed in the calculation does not act on the floe's entire surface and should be less than the maximal speed of individual wind streams.

We can recommend (after Kuznetsov):

for Ω = 10   40   100,000 m^2

w = 34   31   27 m/sec.

In addition, it should be kept in mind that, proceeding from the conditions of ice breakup (disintegration), pressure P_H must not be greater than the critical value.

FIFTH PART

NATURAL METHODS OF DETERMINING ICE PRESSURE ON
STRUCTURAL SUPPORTS

Chapter XVII

EXISTING METHODS OF DETERMINATION

1. Introductory Remarks

The methods reviewed above for determining the pressure
of ice on structural supports are based on extensive experimental material and theoretical concepts, in general corresponding
to the physical pattern of the phenomenon.

Out of necessity, however, we have generalized the phenomenon somewhat, specifically:

a) we have discussed piers of a triangular or semicircular
outline in plane, whereas we find piers of other, more complex
configuration;

b) we have only roughly taken into account the frictional
forces developing between the pier and the floe;

c) we have described only roughly the process of the break
down of a large ice field; and

d) in the recommendation for the working values of the
ultimate strength, we have proceeded in part from the experiments
of samples taken from water and hence existing under somewhat
different conditions.

It is therefore extremely desirable to undertake determina
tions of ice pressure in nature, which will permit us to refine
the working values of the ultimate strength of ice and to introduce certain corrections into the suggestion calculation techniques. In particular, we could conduct an overall calculation
of the effect of various factors (duration of impact, actual
strength of ice sheet, frictional forces form and material of
pier, deformation of floe and structure) without detailing the
frequently quite complex pattern of the phenomenon.

Let us examine the suggestions that have been made along these lines.

a) The eminent specialist in the field of the material mechanics, N.N. Davidenkov [181] proposed undertaking experiments on measuring ice pressure and he designed a device for measuring the static pressure of an ice pack.

The device is a solid metal box, built into a structure, wherein the ice pressure is transmitted to the lid resting on slide-wire ring-type dynamometers. Let us note that such dynamometers were designed later by V.V. Kuznetsov, B.P. Panov and O.K. Vavilova [182]. N.N. Davidenkov suggested that, equating the kinetic energy of the impact to the structure's potential energy, it is possible to obtain results which are close to reality.

For verifying the suggested procedure, Davidenkov proposed building a "device for measuring the kinetic force from a frontal impact of floes" in the form of a powerful buffer with a strong spring mounted on rigid pile sections and he also suggested creating a "device for measuring stresses in the elements of an ice apron" in an experimental structure.

b) B.V. Zylev in his report [45] expressed the opinion that the actual pressure of ice on the piers of bridges with a sloped ice-cutting face can be determined with the aid of ice-breakers.

However, it should be kept in mind that the form of the icebreaker's bow usually differs considerably from that of the supports of bridges and hydraulic engineering structures and, moreover, the frictional forces will be different in both cases.

c) N.N. Petrunichev [48] considers that for taking into account the maximum pressure from a large moving ice field, it is necessary to have a plane of large dimensions through which the force will be transmitted to a recording device. Since such a solution is difficult, Petrunichev analyzes the possibility of modelling the processes of demolishing the ice and of the introduction by such a method of a correcting factor into the theoretical formulas.

At the same time, Petrunichev supports the suggestion concerning the application of electric strain gauges and of packaged sensors based on Brinell's principle.

d) M.E. Plakida [183] considers it necessary to determine whenever possible the values of ice pressure based on structural deformations.

e) The author [57] suggested in 1949 that a determination be made of the pressure of ice pack on structural supports based on the kinematics of the phenomenon.

A large ice field with area Ω moves with velocity v_{Π}, differing somewhat from the surface current velocity v_0. At contact with a pier, the ice field begins to be cut by it and this naturally is accompanied by a reduction in the floe's velocity. The greater the reaction P of support, the more quickly the decrease in the floe's velocity.

The analysis presented below of the conditions of motion of a large floe being cut by a pier provided the opportunity to derive equation linking the reaction of a pier, floe velocity, current velocity, dimensions of floe, its mass, roughness of lower surface of ice, length of ice field sector being cut, with the time spent for cutting.

Using these equations, we can determine the ice pressure in the case of cutting a large floe by a pier. The technique can be called kinematic. Having measured in nature the velocity of a large floe before and after encounter with a pier and also the time (t), having elapsed after the start of cutting, from the expressions derived below it is possible to find the force of pressure.

The suggested method for determining the ice pressure under natural conditions is quite simple and it permits us to establish without complex equipment the actual pressure of ice on a support of various dimensions, form and material.

The method suggested was utilized extensively by us under the conditions of Siberia from 1954-1961 and is explained in detail below.

2. Survey of Determinations of Ice Pressure Under Natural Conditions

In 1946, A.I. Gamayunov determined the amount of ice pressure on bridge supports across the Dnepr R. at Kiev [46] and later also on the Velikaya R. at Pskov.

A. I. Gamayunov made a metal sluice gate with a dimension of 248 X 595 cm, consisting of three hinged connected parts, encircling the cutting face of the pier. From a design standpoint, the gate was made in the form of a frame of No. 24 channel bars and ribs 200 X 12 mm, covered with a sheathing made of 10 mm sheet steel.

During the pressure of an ice field on the hinged suspended gate, pressure on the pier was absorbed by 28 hydraulic dynamometers located between the gate and the supports of the second frame fastened by bolts to the cutting face of the starling.

Ice pressure was transmitted to the cylinder walls of the dynamometers and, deforming them, forced the water (having filled the dynamometers) into metal pipes connected to calibrated piezometers, based on which one could determine the total pressure for the entire group of dynamometers (the bow and two side ones). Prior to the performance of the test operations, all the dynamometers were calibrated in a laboratory and the relationship was established between the amount of pressure and the height of water rise in a glass tube. The overall view of the device is shown in Fig. 55. Thickness of ice during the ice passage varied from 30 to 40 cm while the floes' velocity was around 1 m/sec.

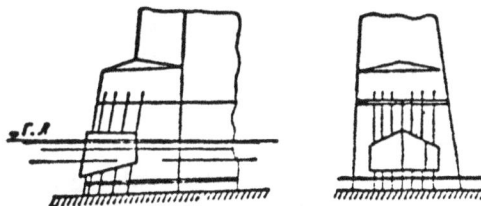

Fig. 55. Determining the Ice Pressure on a Bridge
Support, Conducted by Gamayunov. Key: 1) ice level.

As a result of the tests, considerable fluctuations in ice strength was recorded. The strong floes yielded a sharp impact, the amount of pressure from weak spongy ice (which was observed chiefly during the 1946 ice passage) gradually increased, whereby the piers cut deeply into the ice. Maximal ice pressure on a support reported during the observations comprised 51.2 tons (from a floe with an area of 2000 m^2). Let us point out that Gamayunov himself indicated the imperfection of the measuring equipment (rubber hoses, visual determination of floe area) and suggests that the pressure value obtained cannot be considered maximal.

Essentially, in respect to Gamayunov's studies, it should be noted that they are of interest, providing a possibility of estimating the order of magnitude for ice pressure on bridge supports under actual ice-out conditions. However, the results obtained should not be overestimated for the following reasons:

1) the place of observations (the Dnepr at Kiev, the Velikaya R. at Pskov) was obviously poorly selected for determining the maximal pressure of ice on bridge piers;

2) during the studies, use was made of inadequately modernized equipment, which distorted the actual value for ice pressure

3) no consideration was given to the deformation of the light steel design of the panel absorbing the pressure;

4) the hinged connection of three adjoining gages is ill-advised, since pressure on one of them involves a twisting of the other ones;

5) the visual determination of the floes' dimensions should be replaced by instrumental or even photogrammetric determination as being more precise; and

6) it is necessary to supplement the studies by a measurement of the initial velocity of floes' movement. This would have permitted a comparison of the derived data with the theoretical methods of calculation.

Experiment in determining ice pressure at the Svir'-Stroy (construction site). N.N. Petrunichev [48] and F.I. Bydin [7] describe an experiment in determining the pressure forces of ice, conducted at the construction site of one of the hydraulic stations of the Svir' R. Between the cribwork and metal sheet pile of the upper cofferdam head, a device was installed, consisting of two steel plates with a small ball pressed between them. Based on the dimensions of the impression on the plates, the idea was to judge the amount of the impact. The value of established pressure did not exceed 3.6 tons.

Experiment in determining the pressure of an ice pack in the U.S. Field observations of static ice pressure were conducted from 1946-1951 by the Bureau of Reclamation of the U.S. Department of Interior on several reservoirs located in the mountains of Colorado [104]. The pressure measuring devices were installed between two plates and were arranged in three points according to the height of the ice pack.

Maximal pressure was obtained at the Eleven Mile Canyon Reservoir (Table 47).

Table 47

Values Typifying the Static Pressure of Ice

Years	Pressure, t/r.m.	Ice thickness, m	Remarks
1947-1948	23.8	0.61	reservoir
1948-1949	20.9	0.51	shores: steep,
1949-1950	29.8	0.51	rocky

From 1950-1951, packaged pickup devices were installed at other reservoirs, yielding the following results: with gradually sloping shores from 5.4 to 8.6 tons/rumming m; and with steep shores, from 14.0 to 25.3 tons/running m.

All the reservoirs are situated under approximately the same climatic conditions. The difference in pressures can be explained in terms of the structures, configuration and steepness of shores.

Other determinations. A.M. Ryabukho [86] analyzed a recent case of damage to a pier under one of the bridges in China which occurred under pressure of a large ice field driven by the wind (in December 1955 at air temperature of -20°C).

Thickness of ice pack was 0.4 m; nature of deformation was the displacement of a solid cylindrical support with a diameter of 4.1 m. Considering the deformation of the support, Ryabukho estimated the pressure at 300 tons or 185 tons/m^2. A similar type of determining the ice pressure based on the deformation of structures had also been conducted earlier by other authors. The data obtained from their findings are presented in Table 48 and are explained in Chapter 3.

The methods having existed until now have not found wide application in practice, since although requiring relatively costly and complex equipment, they still did not provide the necessary accuracy. We suggest the kinematic methods substantiated in detail later on. From 1954-1961, this technique was

Table 48

Composite Data on Determining Ice Pressure

Region of world	:Year :	Type of structure :	Kind of damage:	Calcula- ted condition	Remark:	Ref.
U.S.	1920	steel bridge	girder torn off	140 t/r.m.*		[81]
USSR	1944	bridge support on high pile grating	demolition of supports under construction	270 tons	on ea.. pier	[69]
USSR	1941	cribwork pier	displaced cribs	100 t 23.8 t/m²	on ea. crib	[82]
U.S.	1899	roundhead dam	displacement along joint	16.7 t/m²		[1] [60]
Canada	—	solid bridge pier	tilting of pier	71.3 t/m²		[1]
China	1955	" " "	shift of pier	74 t/r.m.* 185 t/m²		[86]

* r.m. = running meter

widely used by the Ice-Thermics Laboratory at the Siberian Branch of the Academy of Sciences of the USSR and by the Novosibirsk Institute of Railway Transport Engineers on the Siberian rivers (Ob', Tom', Tenisey, Angara), characterized by a strong ice passage and representing particular interest.

In all, 40 determinations were made, which exceed greatly the total number of all observations conducted until now in the country by other authors. Part 6 of this book has been dedicated to the research results.

3. Certain Conclusions

1. The determination of actual ice pressure on structural supports is quite desirable, since this would permit an evaluation of the accuracy of the theoretical methods of calculation.

2. It seems that the dynamometers can be utilized with success in determining the static ice pressure. Under winter conditions, an abrupt fluctuation in water levels does not occur often, in connection with which, placing the dynamometers at fixed intervals along the length of the structure's front, it

possible to determine the pressure under various conditions (thickness, strength, temperature of ice, etc.).

3. For finding the dynamic pressure of ice, the dynamometers and measuring cells are much less convenient and require the incorporation of a number of modifications in the equipment design and the observation technique.

4. In connection with this, it is simpler to determine the total pressure of a floe on a structure based on the kinematic pattern of movement of a large ice field, which is being cut by one or several structural supports.

5. Such an approach will permit us to find the actual maximal pressure originating during the cutting of a large floe by a support of given form, dimensions and material.

Chapter XVIII

RATE OF FLOES' MOTION UP TO TIME OF ENCOUNTER WITH
ABUTMENTS

1. Original Concepts

For determining the pressure of an ice sheet on structures, in addition to the data concerning the actual strength of ice, it is also necessary to have information concerning the dimensions of ice fields and the rate of their movement. The calculated reserve of a floe's kinetic energy will permit us to approach·an explanation of the nature of demolition of an ice pack and the forces of interaction developing at this time between the floe and the pier.

Obviously, the velocity of floes will depend on the flow velocity, dimensions and form of floes, forces and direction of wind and also on the extent of coverage of river surface with ice.

Prior to the beginning of ice-out, the flow acts on the floe with force R which is being established with the equation

$$R = f\Omega u^2 = f\, \Omega v_o^2 \tag{18.1}$$

where f = the coefficient of total flow resistance to floe movement; Ω = area of floe; u = relative velocity of flow and floe; v_o = current speed in surface layers of flow.

With an increase in water output and current velocity v_o, force R also increases and finally becomes adequate for breaking off the large ice fields from the stationary ice pack. The separated ice fields under the influence of force R begin to increase the velocity v_{π} of their motion, in connection with which the relative velocity decreases:

$$u - v_{\pi} - v_o \tag{18.2}$$

and hence also the force R of attraction of the floe by the current. Later on, the velocity of floe's movement v_{π} can prove equal to the flow velocity v_o, wherein at this time the force of attraction will obviously equal zero.

If the velocity of floe's movement proves to be greater
than the flow's velocity, the force of attraction

$$R = f \Omega (v_o - v_{JT})^2 \qquad (18.3)$$

already becomes the force of resistance

$$R = f \Omega (v_{JT} - v_o)^2, \qquad (18.4)$$

restraining the movement of the floe.

Thus, the study of the process of an ice field's movement
requires the consideration: a) of floes' dimensions in the initial period of ice passage Ω; b) of current velocity v_o of surface flow layers; c) of coefficient of total flow resistance to
floe's movement; and d) of initial velocity v_{JT} of floe's movement up to the time of encounter with the piers.

While the data listed under points "a" and "b" can be obtained from studying the ice conditions of a waterway, the determination of the other data presents a more complex problem, not
finally solved until recent times.

It should be noted that until now, we have not touched on
the wind's role, which obviously can change the velocity of an
ice field caused by the current.

Since the direction and force of wind are usually subject
to considerable fluctuations, not susceptible to analysis in a
general form, in the subsequent discussion, the wind force is
overlooked. However, in case of the necessity of considering its
influence, the technique discussed below can be utilized.

2. Relative Travel Speed of a Separately Drifting Floe

It is common knowledge that the velocity of a drifting body
differs from the velocity of the fluid particles surrounding it.

As is suggested by V.M. Makkaveyev and I.M. Konovalov [184,
185], such a situation is a result of the fact that in this case,
the lift is not directed vertically (as in the case of a liquid
at rest), but perpendicularly to the free flow surface.

In this manner, if a floe is drifting "separately" (not
touching the shores, bottom, structures or other floes), it is
affected by the following forces indicated in Fig. 56:

a) the actual weight applied at the CG (center of gravity) of the floe (point C), equalling

$$G = \Omega h \gamma_\pi ; \qquad (18.5)$$

b) lift (lifting force), applied to the pressure center (point P) equalling

$$P_o = G \cos \theta , \qquad (18.6)$$

where θ = the inclination angle of free surface of flow to the horizon.

Since the lift P compensates only part of the force of floe's weight G_π the other component

$$G = \sin \theta \approx GI \qquad (18.7)$$

imparts to the floe an acceleration which also leads to the appearance of floe velocity v_π , greater than the flow velocity v_o. Here I = the slope of the flow's free surface.

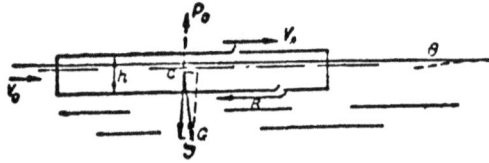

Fig. 56. Forces Acting on a Floe Drifting in a Current.

The relative velocity of floe at this time will equal

$$u = v_\pi \quad v_o , \qquad (18.8)$$

and the force of water's resistance to the floe's motion will be found with the equation

$$R = f \Omega u^2 . \qquad (18.9)$$

Finally, equilibrium can set in between the work of the resistance forces and the work of gravity, after which the floe will move smoothly with a velocity

$$v = v_o + u . \qquad (18.10)$$

The formula of a floe's movement in a general case can be written:

$$\frac{M\,du}{dt} = GI - f\Omega u^2,$$

(18.11)

and in a specific case of uniform movement (u = const) acquires the form:

$$GI - \rho \Omega u^2 = 0$$

(18.12)

from which the relative velocity

$$u = \sqrt{\frac{GI}{f\Omega}} = \sqrt{\frac{\gamma_1 hl}{f}}.$$

(18.13)

3. Factor of Overall Water Resistance to Floe's Movement

It is known that the coefficient of overall water resistance f to the movement of bodies is not constant but rather depends on many factors.

The force of a flow's resistance to the motion of a body can be assumed to equal

$$R = (f_1 S + \varphi \omega)\, u^2,$$

(18.14)

where S = wetted surface of body, m^2 (in the given case, S = Ω); f_1 = coefficient of frictional resistance $Kg/sec^2/m^4$; ω = the submerged part of the body's center ($\omega \approx 0.9\, Bh\ m^2$); φ = coefficient of form's resistance.

Having in mind Eq. (18.9) and (18.14), we derive in a general form

$$R = (f_1 \Omega + 0.9\,\varphi\, Bh)\, u^2,$$

while for a rectangular floe (Ω = BL), we find

$$f = f_1 + 0.9\, h\varphi /L.$$

(18.15)

The problem thus reduces to finding the frictional coefficient f_1 and the form's resistance f.

Let us examine the problem in more detail.

Roughness of ice pack. The frictional factor f_1 is closely linked with the roughness of the ice pack in the pre-spring time which can be quite diversified. As is known, on the roughness of ice, influence is exerted by the conditions of autumn ice passage (presence of underwater hummocks, sludge under the ice), the time having elapsed after the beginning of the ice jamming, the thickness of ice pack, sizes of floe and certain other factors.

The ice pack has the greatest roughness at the beginning of winter while toward its end, the accumulations of frazil ice and hummocks are levelled out by the current. By the beginning of ice-out, the process of levelling the roughnesses converts to the process of eroding the lower ice surface.

In addition, the variations in current velocity across the width of river, plus the uneven thickness of the snow pack, also cause a varying thickness of the ice pack, which in turn leads to its increased roughness.

As was indicated by P.N. Belokon' [186] and by the author [150], in the pre-ice passage time, the values for the coefficient of ice roughness in a river fluctuated from 0.010 to 0.053, where the lower limite pertained to the sectors containing smooth ice, while the upper limit related to the sectors with the presence of sludge under the ice.

Coefficient of frictional resistance for large floes. In connection with the fact that the lower surface of large ice floes in the spring period has significant roughnesses of the order of 0.15-0.30 m and even more (up to 0.5m), their roughness should never be disregarded and a special allowance for it is necessary.

Studying the resistance of water to the motion of rough plates G. Shlikhting [189] proposed for the coefficient f_1 the interpolation formula:

$$f_1 = \frac{\rho}{2}\left[1,89 + 1,62\lg\left(\frac{x}{\varepsilon}\right)\right]^{-2,5},\tag{18.16}$$

where x = length of plate; and ε = absolute value of projections of roughness.

The detailed analysis [150] conducted by us indicated the possibility of using Eq. (18.16) to the case of the movement of ice fields at length x = 50-300 m with roughness projections ε = 0.05 - 0.30 m.

As was indicated below, it can be replaced by a simpler formula:

$$f = \frac{0.0009}{\sqrt[7]{l}} \frac{t \cdot sec^2}{m^4},$$

(18.17)

yielding, however, results which are satisfactory in accuracy, as is evident from Table 49.

A comparison of the calculations based on Eq. (18.17) with observations of resistance to the movement of rafts conducted by the Central Scientific-Research Institute of Timber Floating [187] demonstrated good results. Such a comparison is possible since the roughness of a timber raft is of the same magnitude as the roughness of large ice floes.

Table 49

Values of f_1 10^5 Based on Eqs. (18.16) and (18.17)

Calculation method	Length of ice fields, m			
	50	100	200	300
After Shlikhting				
at $\mathcal{E} = 0.05$ m	43	--	--	--
$\mathcal{E} = 0.10$ m	53	43	--	--
$\mathcal{E} = 0.15$ m	--	49	40	--
$\mathcal{E} = 0.20$ m	--	--	43	38
$\mathcal{E} = 0.30$ m	--	--	--	43
According to $f = \dfrac{0.0009}{\sqrt[7]{l}}$	52	46	41	40

Coefficient of form's resistance. The resistance of form or the drag (frontal resistance) to the movement of a body on the part of a current is often estimated by a formula of the form:

$$R = \varphi \omega u^2,$$

where φ = the coefficient of form's resistance.

Based on the studies made by B.B. Zvonkov [188] and the Central Scientific-Research Institute of Timber Floating [187], for the self-floating timber rafts of rectangular section, it is possible to assume the value:

$$\varphi = 50 \frac{Kg/sec^2}{m^4} = 0.050 \frac{t \cdot sec^2}{m^4}. \tag{18.18}$$

Overall resistance factor. As was explained previously, the overall resistance factor can be found with the formula

$$f = f_1 + \frac{0.9\,\varphi\,h}{l}$$

which in our case acquires the form:

$$f = \frac{0.0009}{\sqrt{l}} + \frac{0.045h}{l} \frac{t \cdot sec^2}{m^4}. \tag{18.19}$$

The calculation of the f-values based on the last formula yields the results shown in Table 50.

Table 50

Values of $f10^5$ Based on Eqs. (18.19) and (18.20)

f - values	h, m	Length of floe, m		
		100	200	300
	1.0	101	69	58
	0.5	73	54	49
	1.0	100	71	58
	0.5	70	50	41

As is evident from the table, the binomial equation (18.19) can be replaced by another

$$f = \frac{\sqrt{h}}{100\sqrt{l}}, \tag{18.20}$$

which coincides satisfactorily with the calculation results based on a more precise equation.

In conclusion, let us point out that since the movement of isolated, separately drifting floes not touching the shores, bottom or adjoining floes does not occur very often (only after the breakup of jams, during scattered ice, at the beginning of ice passage), a study of the actual velocity of floes during the full-scale ice passage is of considerable practical interest.

It should be anticipated that the movement of large floes surrounded by small ones, in the presence of small open spaces is generally slightly restrained. The small floes with a low reserve of kinetic energy and also the large fields, in the absence of open places, can alter their velocity greatly and move slower than the current. We observed the movement of 34 large floes on Siberian rivers during the ice passages from 1950-1956. These natural observations indicated that the velocity of floes in the period of full-scale ice passage as a rule is less than the velocity of the surface current layers and comprises 0.8-0.9 of the latter. However, fluctuations in the ratio $v_\text{л} : v_0$ from 0.70 to 1.07 are quite possible.

4. Effect of Wind on Drift Rate of Floes

The movement of floes during the spring ice debacle is complicated by the effect of wind on the upper surface of the ice pack, usually more or less rough. Obviously, a tail wind is capable of increasing the velocity, while a head wind is capable of reducing the velocity of floes' movement and by the same token of influencing the nature of its breakdown and the value of the developing interaction forces. Wind effect is particularly significant in the large water basins where the movement of floes is essentially determined by the wind influence.

Thus the question reduces to finding the velocity of an ice field with area Ω, depth h under the effect of a wind having a velocity w m/sec, on the upper rough surface of a floe. Academician V.V. Shuleykin [190] as a result of studying the question of the drift of sea ice and natural observations recommends assuming the velocity of floes under the effect of wind propulsion $v'_\text{л}$ based on the formula

$$v'_\text{л} = 0.036\,w \text{ m/sec} \tag{18.21}$$

N.N. Zubov [83] and P.A. Kuznetsov [47] indicate a decrease in the floes' velocity in the littoral zone under the effect of fractures of the shoreline and they recommend assuming

$$v'_\text{л} = 0.02\,w \text{ m/sec,} \tag{18.22}$$

which of course approaches more closely to the river conditions of interest to us.

In determining the possible wind speed w, it is necessary to consider that the calculated wind velocity influences the entire surface of the ice pack and should be less than the maximal speed of individual wind streams (jets). The greater the floes' area, obviously the lower the calculated wind speed should be.

For approximate calculations, we can utilize the recommendations made by P.A. Kuznetsov [47], proposing that we assume w in dependence on the area Ω of floe (Table 51).

Table 51

Calculated w-Values

Ω	10	40	250	1000	thous.of m^2
w	34	31	27	24	m/sec

Since under the conditions of spring ice-out in the large USSR rivers, the area of individual ice fields usually does not surpass 40-100,000 m^2, it is evident that the velocity of floes under wind effect can reach the value:

$$v'_{\Pi} = 0.02 \cdot 31 = 0.61 \text{ m/sec},$$

while for the small floes, it can even be larger.

5. Brief Conclusions

1. The velocity v_{Π} of a separately drifting floe is greater than the velocity v_0 of the flow part surrounding it and can be calculated with the formula:

$$v_{\pi} = v_0 + \sqrt{\frac{GI}{f\Omega}}.$$

2. The value for the overall flow resistance factor can be found with the formula:

$$f = \frac{0,0009}{\sqrt[7]{l}} + \frac{0,045h}{l} \quad \frac{\text{t.sec}^2}{\text{m}^4}$$

or approximately (at $l < 300$ m)

$$f = \frac{\sqrt{h}}{100\sqrt{l}} \; \frac{t.sec^2}{m^4}.$$

3. An analysis of the formulas derived permits us to conclude that the ratio $v_л/v_o$ will usually fluctuate within the limits of 1.10-1.30, decreasing with an increase in depth and a decrease in the floe's area.

4. The greater the flow depth, the less the relative velocity of floe, which can be explained in terms of the decreased relative roughness of floe,(ratio of unevennesses of lower floe surface to the flow depth).

5. In case of the saturation of river surface by ice, the movement of floes is confined owing to friction on shore and the contiguous ice fields. In this case, the velocity of floes (based on our natural [field] observations) comprises 0.8-0.9 of the velocity of the surface flow layers.

Chapter XIX

ANALYSIS OF MOVEMENT OF LARGE FLOES AFTER CONTACTING
WITH STRUCTURAL ABUTMENTS

1. Diagram of Occurrence

During encounter with a structure, to the forces which
have previously acted on a floe, there is added the braking effect of one or several supports, naturally causing a decrease
in the floe's movement. At this time, the stoppage of the small
floes and the splitting of the medium-sized floes is possible.
However, the large floes, having a considerable reserve of
kinetic energy, continue to move, to be sure at a decreased rate.

As is evident from the diagram (Fig. 57), the movement of
a floe can be divided into six stages:

First stage - prior to encounter with support, when the
floe is in steady movement with the velocity:

$$v_{\text{л}} = v_0 + \sqrt{\frac{h\gamma_{\text{л}} l}{f}},$$ (19.1)

caused by its dimensions, roughness and flow gradient.

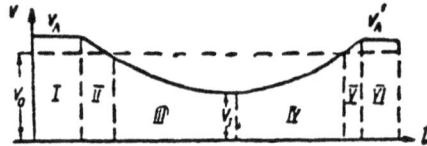

Fig. 57. Stages of Movement of an Ice Field at
Contact with Support.

The second stage develops after the meeting of a floe with
a structural support, when the momentary velocity v of the floe
decreases but is still greater than v_0. At this stage, the
resistance to movement:

$$R = f \Omega (v-v_0)^2$$ (19.2)

also decreases, while at the end of the stage it becomes equal
to zero.

The third stage develops at a further decrease in floe velocity when v becomes less than current velocity v_0. In this case, the floe begins to move slower than the current, experiencing from its side a force of attraction equalling

$$R = \jmath \Omega (v_0 - v)^2 \qquad (19.3)$$

and constantly increasing in the course of time.

Starting from the third stage, force R will transmit to the ice floe additional energy and thereby promote its breakdown. As calculations indicate, the additional energy received from friction on the lower rough surface of an ice floe can play a very significant part in the pattern of the demolition processes, comprising up to 80% of the initial reserve of kinetic energy.

Fourth stage. At the time of completion of an ice floe's demolition, its momentary velocity becomes equal to v_1, after which the remnants of the floe cut into parts enter the stage "acceleration", moving with ever-increasing velocity.

The fifth stage develops at a further increase in the velocity of the cut parts of the ice floe to the values higher than v_0.

The sixth stage is characterized by the newly established uniform movement of parts of the cut-off ice floe.

It should be kept in mind that in case of an inadequate reserve of kinetic energy of a floe, the stoppage of floes is possible at the piers during the second or third stages.

2. Conditions of Ice Field's Movement

(First stage--uniform movement of ice floe prior to its contact with a structure)

Having examined the system of effective forces presented in Fig. 57, let us write an equation for the motion of an ice floe in the form:

$$\frac{\mathcal{M} dv}{dt} = GI - R = GI - \jmath \Omega u_1^2 . \qquad (19.4)$$

Since, at uniform movement, velocity is constant, from the formula we find

$$u_1 = \sqrt{\frac{GI}{\jmath \Omega}}$$

and

$$v_{\text{п}} = v_0 + \sqrt{\frac{GI}{f\Omega}}$$

in a form already known to us.

It is obvious that in the ensuing stages, the formula for the ice floe's motion acquires the form (Fig. 58)

$$\frac{\text{м} dv}{dt} = GI \pm R - P. \tag{19.5}$$

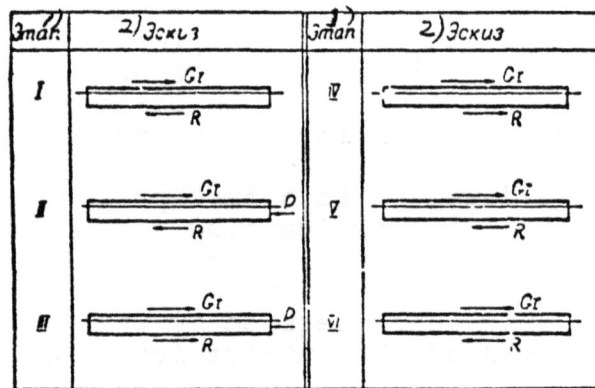

Fig. 58. Diagrams of Forces Acting on an Ice Floe at Different Stages. Key: 1) stage; 2) sketch.

Having reviewed this formula, we can obtain a relationship for the velocity of ice floe at the moment:

$$v = f(t, v_{\text{п}}, \text{м}, P, f\Omega, v_0); \tag{19.6}$$

the time of its passage of distance x -

$$t = f(v_0, v_{\text{п}}, \text{м}, f\Omega, GI, P); \tag{19.7}$$

the path s, covered by an ice floe during time t, -

$$s = f(v_0, v, \text{м}, t, f\Omega, GI, P). \tag{19.8}$$

Detailed conclusions from the formulas have been published in report [150], so that we will limit ourselves here to listing only the combined data (Table 52).

In the table shown, we have introduced the following notations: v_0 = velocity in m/sec of surface flow layers; $v_Л$ = velocity of ice floe in first stage, m/sec; v = velocity of ice floe at moment t, m/sec; m = mass of ice floe, t sec^2/m; Ω = area in m^2 of ice floe; $c = P + f\Omega (v_Л - v_0)^2$; h = thickness in m of ice floe; $b = f\Omega t \cdot sec^2/m^2$; f = coefficient of general resistance to ice floe movement, t sec^2/m^4; P = force of ice pressure, t, on support; t = considered moment of ice floe's movement; and t_2 = time necessary, seconds, for passage of ice floe in the second stage.

Table 52

Working Formulas Determining the Conditions of Ice Floe's Movement in a Current

Data	first stage	second stage	third stage
Velocity of floe's movement	$v_л = v_0 + \sqrt{\dfrac{\tau_л hl}{f}}$	$v = v_л - \dfrac{ct}{M}$	$v = v_0 - \sqrt{\dfrac{c}{b}}\,\text{th}\,\dfrac{(t-t_0)\sqrt{bc}}{M}$
Time of stage's passing		$t_2 = \dfrac{M}{c}(v_л - v_0)$	$t_3 = t_2 + \dfrac{M}{\sqrt{bc}}\,\text{ar th}\left[(v_0-v_1)\sqrt{\dfrac{b}{c}}\right]$
Path covered in stage		$S_2 = \dfrac{M}{2c}(v_л^2 - v_0^2)$	$S_3 = S_2 + v_0(t-t_2) - \dfrac{M}{2b}\ln\dfrac{c - b(v_0-v)^2}{c}$

Thus, having analyzed the conditions of the movement of an ice floe which is being cut by a structural support, we derived a series of formulas:

$$v = f(t) \text{ and } S = f(v),$$

permitting us to solve the practical problems primarily connected with finding the pressure of ice on the structural supports.

3. Suggestions Made by F. I. Bydin

In his lecture presented at the Third All-Union Hydrologic Conference, F. I. Bydin [191] proposed a formula for the rate of ice cutting by structural supports in the form:

$$v_{pe3} = \frac{v_{cB}}{1 + \left(\frac{h}{A}\right)^n} = \frac{v_{cB}}{1 + \left(\frac{h}{30}\right)^2},$$

(19.9)

where v_{pe3} = rate of ice cutting by supports, m/sec; v_{cB} = rate of natural ice movement, m/sec; and h = ice thickness, cm.

The formula reflects properly the reduction in the velocity of ice floe after contact with an obstacle but it stresses its dependence only on the depth of the ice pack.

It appears that the decrease in velocity depends above all on the mass of ice field and v_{pe3} can not be constant. Incidentally, the necessity of a variable value for the A-factor was indicated by Bydin personally.

4. Analysis of Conditions of Floe's Movement

The analysis of the equations derived is accomplished most simply with a numerical example, illustrating the conditions of movement of large floes after their contact with the structural supports.

Let us adduce one example (of the 17 performed by us) for an ice floe being cut by a pier.

Example of calculation. An ice floe of given dimensions and strength, having moved at a fixed velocity, has encountered a vertical bridge support. Find the conditions of further movement of the ice floe. The initial (raw) data for the solution of this problem are listed in Table 53.

Conditions of ice floe's movement prior to encounter with a structure. In the first stage, a separately floating ice floe exists under conditions of uniform movement with a velocity which is found with the formula:

$$v_{\scriptstyle 2} = v_0 \left(1 + \frac{8n\sqrt[4]{hL}}{H_0^{2/3}}\right) = 2{,}46 \quad \text{m/sec}$$

obtained by us with the utilization of Chézy's formula [150].

-232-

Table 53

Initial Data

Parameters	Notation	Value	Unit of measurement
Area of ice floe	Ω	100 000	m^2
Size of ice floe	L,B,h	316·316·1	m
Average depth of flow	H_o	7.0	m
Surface current velocity	v_o	2.0	m/sec
Coefficient of river-bed's roughness	n	0.025	--
Width of pier	b_o	3.0	m
Ultimate crushing strength of ice	R_C	65.0	t/m^2
Coefficient of non-dense osculation	k	0.80	--

Remark. 1. The support is solid and the influence of its deformation can be disregarded. The coefficient of pier's form m = 1.
2. Impact is centralized.

Conditions of ice floe's movement during the second stage. The additional calculations:

a) mass of ice floe

$$M = \frac{\Omega h \gamma_{\pi}}{g} = 9378 \ \frac{t/sec^2}{m};$$

b) coefficient of total resistance of flow:

$$f = \frac{\sqrt{h}}{100\sqrt{L}} = 0,00056 \ \frac{t \cdot sec^2}{m^4};$$

c) we calculate the ice pressure on a solid support, disregarding its strains but with allowance for the nondense osculation factor based on the formula:

$$P = b_0 hk R_{cм} m = 3 \cdot 1 \cdot 0{,}80 \cdot 65 \cdot 1 = 150 \ m;$$

$$\text{г)} \ b = f\Omega = 56 \frac{t.\sec^2}{m^2};$$

$$c = P - f\Omega (v_л - v_0)^2 = 150 - 56 (2{,}46 - 2{,}00) = 138{,}2 \ m;$$

$$\sqrt{bc} = 88 \frac{t.\sec}{m};$$

$$\sqrt{\frac{b}{c}} = 0{,}636 \frac{\sec}{m};$$

$$\sqrt{\frac{c}{b}} = 1{,}57 \frac{m}{\sec}$$

We calculate the passage time of an ice floe through a stage based on the precise formula

$$t_2 = \frac{м}{\sqrt{bc}} \ \text{arctg} \left[(v_л - v_0) \sqrt{\frac{b}{c}} \right] = 30{,}4 \ \sec$$

or based on the approximate formula

$$t_2 = \frac{м}{c} (v_л - v_0) = \frac{9378}{138{,}2} 0{,}46 = 31{,}2 \ \sec;$$

yielding a deviation of only 2.6%.

According to the precise formula, the path covered during the stage equals

$$s_2 = v_0 t_2 + \frac{м}{2b} \ln \frac{P}{c} = 2 \cdot 30{,}4 + \frac{9378}{112} \ln \frac{150}{138{,}2} = 67{,}7 \ m,$$

based on the simplified formula

$$s_2 = \frac{м}{2c} (v_л^2 - v_0^2) = \frac{9378}{276{,}4} (6{,}05 - 4{,}00)^2 = 69{,}5 \ m,$$

also yielding a deviation of only 2.6%.

The velocity at various time moments is found with the equation

$$v = v_{\Pi} - ct/m.$$

Let us conduct the calculations of v

at t =	0	10	20	30.4 sec
v =	2.46	2.31	2.16	2.00 m/sec

As was to be expected, at the end of the stage, velocity equals v_0 which can serve as a control over the correctness of the calculation.

The path covered in various time moments will be found with the equation:

$$s = \frac{M}{2c}(v_{\Pi}^2 - v_0^2) \approx \frac{v_{\Pi} + v}{2} \cdot t$$

at t =	0	10	20	30.4 sec
s =	0	23.8	46.2	67.7 m,

i.e. the velocity of the ice floe's movement becomes equal to the flow velocity only after the support has cut into it for the distance of 67.7 m.

Conditions of movement in the third stage. The theoretical value of the final velocity is found based on [150]:

$$v_1 = v_0 - \sqrt{\frac{c}{b}} = 2,00 - 1,57 = 0,43 \text{ m/sec}$$

The time required for reaching velocity v'$_1$ can be found based on [150];

$$t_3' = t_2 + \frac{M}{\sqrt{bc}} \text{ arth} \left[(v_0 - v_1) \sqrt{\frac{b}{c}} \right] = \infty.$$

Let us note that the velocity of ice floe approaches v'$_1$ only asymptotically, never reaching it.

Therefore, in practice the final velocity of an ice floe at the end of a stage can be determined, if only in the following manner:

$$v_1 = 1.05, \quad v'_1 = 0.45 \text{ m/sec.}$$

The time necessary for reducing the velocity of 0.45 m/sec can be computed on the basis of Table 52.

$$t_3 = 30,4 + \frac{9378}{88} \operatorname{arth}[(2 - 0,45)\,0,636] = 295 \ \text{sec}$$

The velocities at various time moments will be found based on:

$$v = v_0 - \sqrt{\frac{c}{b}} \operatorname{th}\left[\frac{(t - t_2)\sqrt{bc}}{\mu}\right] = 2 - 1,57\operatorname{th}(0,094t - 0,286).$$

Let us present the calculations of v

at t =	30.4	50	100	150	200	250	295 sec
v =	2.00	1.72	1.10	0.73	0.56	0.48	0.45 m/sec

The distance covered in various time moments is determined as:

$$s = s_2 + v_0(t - t_2) - \frac{\mu}{2b} \ln \frac{c - b(v_0 - v)^2}{c}.$$

Let us conduct the calculation at the given relationship:

at t =	30.4	50	100	150	200	250	295 sec
s =	67.7	103.9	173.9	216.9	253.9	279.9	287.9 m

As we shall observe, the ice floe reaches a velocity of 0.45 m/sec when the support has cut into the ice only for a distance of 287.9 m, less than the floe length L = 316 m. However, the time necessary for the total cutting through the ice floe is found from Table 52 at s = L.

$$t = \frac{1}{2}\left[L - 6,9 - 83,6 \ln \frac{138,2 - 56(2 - 0,45)^2}{138,2}\right] = 357 \ \text{sec}$$

The conditions of ice floe motion in the fourth and fifth stages when parts of the cut-off floe enter the stage of "acceleration" do not have practical significance for the operation of the structure and therefore are not determined by us.

The nature of an ice floe's movement is illustrated in Fig. 59 which reflects the law of decrease in velocity of a moving ice floe.

As we shall observe, in the second stage (and at the beginning of the third), the ice floe velocity drops almost according to a linear law and only at the attainment of a velocity equalling $\approx$ 1.5 of the final velocity v_1 does the law of variation in v become modified abruptly.

As has been indicated below, such a nature of variation is associated with the increase in the force of the ice's attraction by the flow, which transmits the additional energy to the floe and by the same token stops the intensive decrease in the rate of its movement.

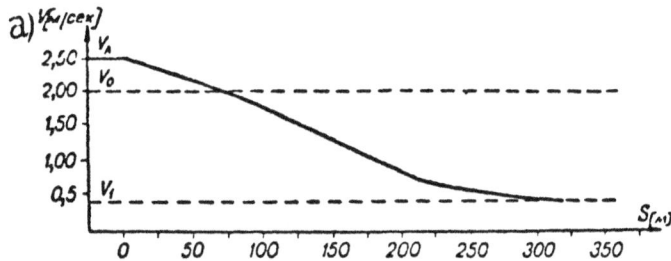

Fig. 59. Nature of Reduction in Velocity of an Ice
Floe After Contacting a Bridge Support, for the
Conditions of the Example Under Review.
Key: a) V [min/sec].

In connection with this, it is interesting to analyze the power balance (budget) of an ice floe for the entire period of its movement; this has been done below.

5. Energy Reserve Required for Complete Cutting
Through of a Floe

An ice field moving with a fixed velocity has the known supply of kinetic energy which upon encounter with a structure, is spent in the work of demolishing the ice floe by the supports. Analyzing the question, it is necessary to take into account that in the process of sliding over the free surface of flow, the ice floe acquires a certain additional force $T_{ук.л}$ owing to the transition of part of the potential energy to kinetic energy.

Finally, it is necessary to take into account the additional energy of the flow developing owing to friction on the lower rough surface of the ice floe during the process of its demolition, i.e. $T_{в.л}$.

In conformity with the law of kinetic energies, we have

$$\frac{m\left(v_{\text{л}}^2 - v_1^2\right)}{2} = T_{\text{укл}} + T_{\text{вл}} + T_\text{p},$$

where m = mass of ice floe; $v_{\text{л}}$, v_1 = initial and final velocity of ice floe; T_p = work expended in demolishing the ice floe; $T_{\text{укл}}$ = the energy developing during the sliding of an ice floe on the surface of flow; and $T_{\text{вл}}$ = the work of frictional forces of flow on the floe's lower surface.

Let us examine all terms in this formula.

a) the initial reserve of the floe's kinetic energy can be found with the formula:

$$T_\text{н} = mv_{\text{л}}^2/2.$$

In the calculation of $T_\text{н}$, it is necessary to take into account the actual velocity of ice floe, which, as we observed above, can be greater by 10-30% than the velocity of the floe's surface layers and can exert a significant influence:

$v_{\text{л}}$ = v_o 1.10 v_o 1.20 v_o 1.30 v_o

$T_\text{н}$ = T   1.21 T   1.44 T   1.69 T

In this manner, the failure to consider the actual velocity of an ice floe's movement can involve an underestimation of its kinetic energy by 20-70%.

b) the final reserve of kinetic energy of an ice floe is determined by the floe's velocity at the time of completion of its demolition:

$$T_\text{k} = mv_1^2/2.$$

c) the force acquired by an ice floe owing to slippage on the surface of flow can be taken into account as follows. If by the moment under consideration, a support has cut into a floe by the distance x meters, for this distance, the component of weight force GI of the ice floe will accomplish the work.

$$T_{\text{укл}} = GI\ x,$$

which also typifies the value of the part of potential energy having converted to a kinetic form. Moreover, during uniform movement, equilibrium should exist between the component force of gravity, and the resistance of flow to the movement of ice floes, i.e.

$$GI = R = f\Omega \ (v_1^2 - v_0^2),$$

therefore the work $T_{yкл}$, we can also suggest another formula:

$$T_{yкл} = f\Omega \ (v_1^2 - v_0^2) x.$$

d) the energy spent for the demolition of an ice floe. The ice floe, having contacted a structure, begins to be demolished by the piers which cut into it for some given distance, depending on the supply of the ice floe's kinetic energy, sizes and shapes of pier, and also the strength of ice.

In this connection, it is obvious that the ice pressure on a structure (in proportion to the penetration of the support into the ice) increases continuously, however only until the support has penetrated for its entire width. Starting from this moment, the ice pressure on a support becomes about constant and dependent on the given actual conditions.

Signifying by P the reaction force of one (or of several) piers, we obtain the total work spent in introducing a support (or supports) for the distance of x meters, in the form $T_p = P\ x$.

e) the energy brought by the flow during the demolition of the ice floe.

It is evident that the energy of the flow, developing from friction on the lower rough ice floe surface can be found from the equation of energy balance for the entire period of the ice floe's demolition.

If the kinetic energy reserve is sufficient for the complete demolition of the floe for the entire length L, the formula:

$$T_н \ + T_{yкл} + T_{эл} = T_p + T_k,$$

is valid, from which we can also find the energy introduced by the flow for the entire period of demolition:

$$T_{эл} = \frac{м(v_1^2 - v_л^2)}{2} + L\ (P - GI).$$

Thus, the minimally required energy supply sufficient for the total breakdown of the ice field equals

$$T_{M} \geqslant T_{p} - T_{y\kappa\pi} - T_{B\pi} + T_{k}.$$

If the energy supply is less, the cutting of a support (or supports) into the ice field for the distance x will occur, after which the ice floe stops.

Hence, for explaining the conditions of ice floes' breakdown, it is necessary to know how to determine at any time the floes' velocity v, depth x of pier's penetration into the ice; one must also be able to calculate the final floes' velocity, v_1.

6. Energy Balance of Large Floes During Their Cutting by Abutments

Let us now discuss the nature of variation in energy reserve of floes in the actual example which has just been examined.

The required expenditure of energy for breaking up the ice floe equals

$$T_{p} = PL = 150 \cdot 316 = 47,400 \text{ tons/m}$$

However, at the time of meeting with the pier, the ice floe had the initial energy reserve of only:

$$T_{H} = mv^2/2 = 9378 \cdot 2.46^2/2 = 28,375 \text{ tons/m},$$

which comprises only 60% of that necessary.

It would appear that the ice floe should have stopped at the structure but nevertheless it was not only cut through by the pier but, toward the end of the process, it even had the final reserve of energy equalling:

$$T_{k} = mv_1^2/2 = 9378 \cdot 0.45^2/2 = 906 \text{ tons/m}.$$

Such a situation is explained by the appearance of other, additional energy sources, both owing to the slipping of the ice floe over the free surface of the river and also especially owing to the tractive force of the floes by the current (flow).

Let us consider the energy balance of ice floes at various time moments. The calculations are summarized in Table 54 and are illustrated by Fig. 60.

Table 54

Energy Balance of Floe's Movement, tons/m

Paths covered and forms of energy	Time, sec, from beginning to breakdown										
	0	10	20	30 4	50	100	150	200	250	295	357
path covered	0	23,8	46,2	87,7	103,9	173,9	216,9	253,9	279,0	287,0	316,0
$T_н$	28375	28375	28375	28375	28375	28375	28375	28375	28375	28375	28375
$T_{укл}$	0	283	548	805	1232	2060	2580	3010	3310	3420	3750
$T_н + T_{укл}$	28375	28658	28923	29180	29607	30435	30955	31385	31685	31795	32125
$T_{разр}$	0	3570	6940	10150	15580	26050	32500	38000	41800	43100	47400
$T_{кон}$	28375	24950	21800	18760	13950	5670	22490	1465	1080	950	906
total energy supply	28375	28520	28740	29910	29480	31720	34990	39465	42880	44050	48300
$T_{вл}$	0	—138	—183	—270	—127	+1285	+4035	+8080	11195	12255	16175
In % of $T_н$	0	—0,4	—0,6	—1,0	—0,4	+4,5	+14	+29	+40	+43	+57

Analyzing the table and Fig. 60, we conclude:

a) after the encounter of the ice field with the structure, the supply of the floe's kinetic energy is spent in the work of demolition and in connection with this, it decreases continuously.

If the supply of kinetic energy of an ice floe proves to be inadequate for cutting the floe through its entire length, the ice field will be stopped by the structural supports.

b) in the process of breakdown, there will be applied to the ice floe the additional energy $T_{укл}$ (grad)' and also $T_{вл}$ (trac).

c) The energy of gradient

$$T_{укл} = \int \Omega (v - v_0)^2 \ x$$

-241-

increases continually in proportion to the pier's cutting into
the ice according to a linear law but it plays a relatively
slight part in the energy balance, even that of large ice floes.

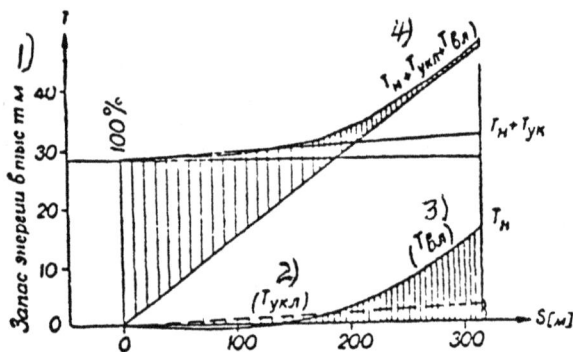

Fig. 60.   Energy Balance of an Ice Floe Being Cut
   Through by a Pier (Support).   Key: 1) energy balance
   in thousands of ton/meters; 2) $T_{gradient}$;
   3) $T_{tractive}$; 4) $T_H + T_{grad} + T_{trac}$.

In our example, the force of gradient comprises from
0 to 13% of the original energy reserve.

d) the tractive force

$$T_{в.л} = (T_H + T_{ук.л}) - (T_{разр} - T_k)$$

in the second stage even decreases the floe's energy supply (to
be sure, trivially, owing to the work of the flow's resistance)
to the movement of the ice floe.   After the arrival at the third
stage, the resistance changes its sign and becomes the tractive
force

$$R_{trac} = f\Omega(v_o - v)^2,$$

initially increasing slightly and then very intensively.

We can consider that the intensive increase in T_{trac}
begins after the pier has cut into the ice floe for the doubled
length of the second stage.

$$L = 2s_2.$$

The influence of T_{trac} on the energy balance can be very considerable. In the example under review, the value attained 57% of T_H, which basically established the nature of the ice floe's breakdown.

e) in the case of the movement of an ice floe with the initial velocity

$$v_1 < v_0$$

(which takes place in the period of full-scale ice passage), the value of T_{trac} should be more considerable, since the energy will be supplied by the current immediately, beginning from the time of contact with the support.

Thus, we should infer that it is inadmissible to consider the conditions of an ice floe's demolition without a detailed energy balance.

7. Necessary Conditions for Stopping the Floes at the Abutments

Let us examine the conditions of ice floes' stopping at a structure. Obviously, the greater the supply of kinetic energy that the given ice field possesses, the fewer the bases for its stopping by the supports.

Let us determine the dimensions of ice floes which are being stopped by the piers and we will also anlayze in detail the question concerning the time required for stopping the large ice floes.

Dimensions of various floes which are stopped by piers. For the ice floes stopped by a pier based on Table 52 the final velocity equals,

$$v = v_0 - \sqrt{\frac{P - Gl}{f\Omega}} = 0,$$

whence

$$f\Omega v_0^2 = P - f\Omega (v_n - v_0)^2.$$

Considering the movement of ice floes in the period of full-scale ice passage (at $v_n = v$), we find the area of the floes which are being halted:

$$\Omega = P / f v_0^2 \tag{19.10}$$

-243-

Assuming in first approximation that $f = 0.00065 = const$, we finally find:

$$\Omega = 1.5P/v_o^2 \quad \text{thousands of m}^2 \qquad (19.11)$$

Let us stress that in the Eqs. (19.10, 19.11), by the Ω-value we connote not only the area of one separately drifting ice floe but also the total area of several ice fields exerting pressure on each other. At this time, the ice floes seemingly combine their efforts and can be considered compositely.

The time of ice floes' stopping equals (at $v_{\pi} = v_o$)

$$t_{oct} = \frac{\varkappa}{\sqrt{bc}} \text{arth} \left(v_0 \sqrt{\frac{b}{c}} \right).$$

Naturally, the slower the current speed, the shorter the stopping time of the ice floe. In connection with this, let us consider the time of floes' stopping, having a current velocity equal to or less than 1 m/sec.

We can demonstrate that in this connection:

$$t_{oct} = \frac{\varkappa v_0}{c} = \frac{\Omega h v_0}{10P}. \qquad (19.12)$$

It is natural that at $v > 1$ m/sec

$$t_{oct} > \frac{\Omega h v_0}{10P}.$$

Calculations based on the last formula are presented in Table 55.

In this manner, the large ice fields require for their stopping a time amounting to several seconds or even tens of seconds. Figure 61 portrays the dependence of t_{stop} upon velocity v_0 and reaction P of support.

Table 55

Time, Seconds, of Ice Floe's Stopping

| h,m | P | v_0 m/sec | area of floe, thous. of m^2 | | | | |
			20	30	50	80	100
1,0	100	0,5	—	14,4	23,8	38,2	47,6
		1,0	19,2	28,6	47,6	76,4	45,4
		2,0	38,4	57,2	48,2	152,8	190,8
1,0	200	0,5	—	—	12	19,2	23,8
		1,0	—	14,4	23,8	38,2	47,5
		2,0	19,2	28,6	47,4	76,4	95,4

-244-

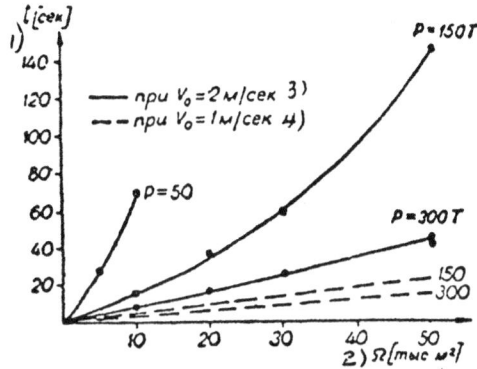

Fig. 61. Time Required for Stopping an Ice Floe.
Key: 1) t [sec]; 2) Ω [thousands of m^2]; 3) at
v_o = 2 m/sec; 4) at v_o = 1 m/sec. v_o = velocity
of floe's movement; p = reaction of pier.

8.   Certain Conclusions

1. A large ice field, having met one or several structural supports, decreases its velocity as a result of the braking
effect from the pier.

2. Initially, the travel speed of the ice floe drops
(practically) according to a linear law and then according to a
hyperbolic law, approaching asymptotically to a certain final
velocity.

3. The variation in the law of velocity reduction (in
proportion to the cutting of the ice floe by the support) is explained by the ever-increasing role of the additional kinetic
energy absorbed by the ice floe.

The additional energy develops both owing to the frictional forces between the flow on the lower rough surfaces of the
ice floes ("tractive force-T_{trac}"), as well as owing to the
slipping of the ice floe along the free surface of the current
("energy of gradient - T_{grad}").

4. A decisive role in the balance of the ice floe's
kinetic energy is played by the "tractive energy" comprising up
to 50-60% of the initial energy supply. The gradient energy
plays a lesser part but it also increases the supply of the
floes' energy by 5-15%.

5.   In connection with this, the consideration of the conditions of an ice floe's disruption without consideration of the

-245-

additional inflow of energy leads to great errors and therefore
is inadmissible.

6. When the final velocity being established on the
basis of Eq. (19.13), proves to be less than zero

$$v_1 = v_0 - \sqrt{\frac{c}{b}} < 0, \qquad\qquad (19.13)$$

this testifies to the possibility of stopping the ice floes at
the structural piers.

7. The large ice floes having a considerable reserve of
kinetic energy overcome the braking effect of the structural
supports more successfully. The maximum area of the ice floes
which have already been stopped can be computed roughly based
on the formula at Pmv_0 m/sec:

$$\Omega \approx 1.5 P/v_0^2 \text{ thousands of } m^2$$

8. The time required for stopping the large floes at the
structural supports is considerable (running from several to 100
and more seconds).

Therefore, it should be considered that there takes place
here not an impact but only the effect of the short-lived action
of the forces.

The stopping time of the large ice floes can be calculated
with the formula

$$t_0 = \frac{M}{\sqrt{bc}} \text{ arth}\left(v_0 \sqrt{\frac{b}{c}}\right).$$

9. At low velocities of an ice floe's movement (v<1 m/sec)
this expression assumes the form:

$$t_0 = \frac{\Omega h v_0}{10 P} \text{ sec}$$

in connection with which we should consider that the relationship

$$t_0 > \frac{\Omega h v_0}{10 P} \text{ sec}$$

also takes place.

10. The conclusions in the present chapter have been developed under the assumption that the force of interaction P between a support and an ice floe does not change during the process of cutting or with the making of several assumptions.

Therefore, it is quite desirable to conduct natural observations of the velocity of ice floes which are being cut through by bridge supports.

Chapter XX

PROPOSED METHOD OF DETERMINING ICE PRESSURE IN NATURE

1. Gist of Suggested Method

An analysis of the movement of an ice floe during its cutting by a pier under a structure permits us to suggest a new method for finding the actual ice pressure in nature during the ice-out period.

As has been demonstrated, a large ice floe of given dimensions Ω, h moves with a velocity v_{Π} and after meeting with a structure, it starts to be cut by the pier and naturally reduces its travel speed according to a fixed law. The curve showing the relationship of the travel speed of the ice floe and the time can vary depending on the mass of ice floe, its initial travel speed, ultimate ice strength, and the dimensions and form of the pier in plane and profile.

Let us comment that the actual process of breakdown of an ice floe has a pulsating nature. The support cuts into the ice fields, the interaction forces increase, and then decrease after a local disruption of the ice floe's edge. Subsequently, the process is repeated cyclically, as this is portrayed e.g. in Fig. 62.

Having such a graph and the data concerning the dimensions of an ice floe, we can compute its mass and acceleration (j), after which it is easy to find the pressure

$$H \approx mj \qquad (20.1)$$

at any time moment.

The method can be called kinematic, since it is based on the measurement of the ice field's velocity after meeting with a structure.

Without using complicated equipment, and with minimal expenses, the kinematic method permits us:

a) to establish the actual resistance of ice to the penetration into it of piers of various form, sizes and material; and

b) to introduce corrections into the Technical Specifications and standards for establishing the ice loads for various climatic regions of the country.

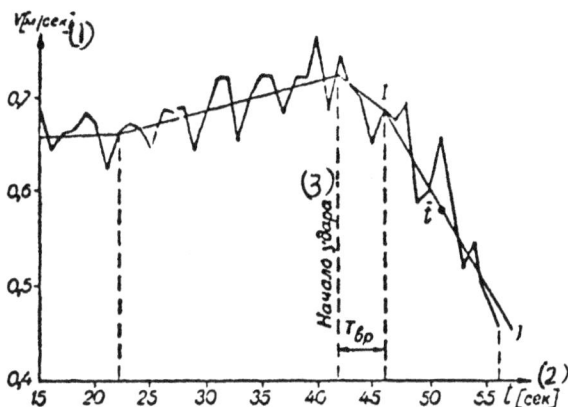

Fig. 62.   Variation in Velocity of Ice Field After Meeting With the Sloped Face of a Pier.
Key: 1) v [m/sec]; 2) t [sec]; and 3) start of impact.

Our methods also permit us to estimate the value of actual ice pressure on the structural supports when it was not possible to construct a curve of the relationship for the velocity of ice floe movement to time.   It is sufficient to establish the dimensions of the ice floe and the rate of its movement at any two time moments.

In connection with this, we suggest two variants in the procedure for finding the ice pressure in nature.

2. First Method of Determining Ice Pressure

(at absence of a curve reflecting the decrease in velocity of ice floe movement during its breakdown)

Necessary Data.   In this case, it is necessary to have information concerning the dimensions of an ice floe and the velocity of its movement during contact with a support and after the time t following the encounter.

Sequence of Calculation. A large ice floe with area Ω, thickness h is moving with velocity v_{Π}, generally speaking differing from the current velocity v_o in the surface layers of flow.

After contact with a structure, the large floe begins to be cut by the pier and to decrease the rate of its movement to a certain value v. An analysis conducted in the preceding chapter has shown that during the time t, the velocity of the ice floe manages to decrease to the value:

$$v = v_{\Pi} - ct/m \tag{20.2}$$

in the second stage, while in the third stage, it drops to the value:

$$v = v_0 - \sqrt{\frac{c}{b}}\, \text{th}\, \frac{t\sqrt{bc}}{\mu}. \tag{20.3}$$

Let us recall that the pressure P of ice is linked with the parameter c in the following manner

$$P = c + f\Omega(v - v_o)^2 \tag{20.4}$$

Thus, having determined in nature the current velocity v_o, the initial and final velocity of ice floe's movement v_{Π}, v, its dimensions Ω, h and having noted the time t, during which the velocity has dropped to v, from Eqs. (20.2) or (20.3), we can ascertain the pressure P.

Since the time of an ice floe's movement in the second stage is short (0.5-20 sec), it is handier to use Eq. (20.3) for the third stage of the floe's movement, solving it for the value c.

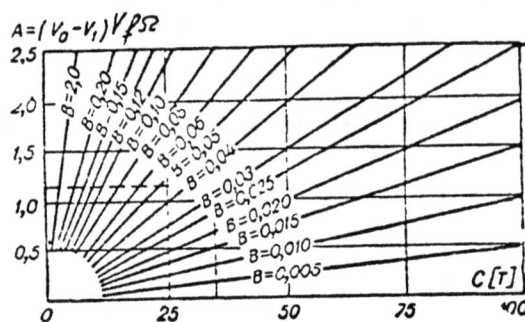

Fig. 63. Graph for Solving Eq. (20.5).

For facilitating a solution of Eq. (20.3), we suggest the graph (Fig. 63). As we observed, the question can be represented in the form:

$$E = \sqrt{c} \ thB \sqrt{c}.$$ (20.5)

where

$$A = (v_o - v) \sqrt{b}$$ (20.6)

and

$$B = t \sqrt{b/m}$$ (20.7)

Having found A and B, directly from the graph in Fig. 63, we can find the c-value and then the unknown ice pressure on a support from the formula:

$$P = c + f\Omega (v_{\sqrt{\eta}} - v_o)^2$$

The procedure for conducting the field operations is explained below, simultaneously for the first and second methods.

3. Second Method of Determining Ice Pressure

(in the presence of a curve for the decrease in velocity during the disruption of an ice floe).

Necessary Data. In this case, it is necessary to have data concerning the floe dimensions Ω, h and concerning the variation in the velocity of its movement during the cutting process.

Sequence of Calculation. Having actually determined in nature the floe dimensions (Ω, h) and the curve of the relation ship of the decrease in its velocity during its cutting by the pier, we find the mass of floes:

$$\mathcal{M} = \frac{\Omega h \gamma_\pi}{g} \simeq \frac{\Omega h}{10,6} \frac{t \cdot sec^2}{m^2}$$

and the maximum acceleration [according to the curve $v = f(t)$] for the time period from t_1 to t_2:

$$j = \frac{v_1 - v_2}{t_2 - t_1} \ m \cdot sec^2$$

after which it is easy to find the ice pressure on a support

$$P = mj$$

The pressure of ice changes during the cutting process, in connection with which it is necessary to select the "steepest" sector of the curve

$$v = f(t),$$

yielding the maximum value of acceleration and consequently of ice pressure on support.

Volume and Nature of Required Observations. We described below the procedure which was used by us.

1. We chose the existing bridge crossing for a spillway dam, if possible with a straight approach of floes perpendicular to the axis of the structure.

2. Several days prior to the beginning of ice passage, we measured the ice thickness at 1-2 sections, at 10-30 points.

3. We set up a movie camera (either type KS-50 or KONVAS) at one of the structural supports.

4, At the approach of a large ice field to an adjoining pier, we made movies of the process of the field's disruption.

5. We conducted detailed observations of the fluctuation in the water level.

6. The width of ice field was measured by two observers standing on the span structure of the bridge above the chunks of the moving ice field.

7. The thickness h of ice pack was found approximately, on the basis of systematic measurements in the period directly preceding the ice passage and also based on photographs of the process of the disruption of ice fields by the pier's cutting face.

A comparison of the thickness (visible during its climbing onto the sloped face of the pier) with the known measurements of the facing rocks provides the possibility of roughly estimating the thickness of the ice pack.

At contact with a vertical support, one can also sometimes establish the thickness of ice floes rising during disruption. Finally, we perform the measurement of thickness of the floes being crowded onshore during the ice passage.

8. The processing of the data was conducted as follows.

The movie film was placed in a projector providing the possibility of obtaining a magnified image of any given frame, in which one could clearly observe the part of the bridge pier and of the ice field. Noting in the frame of the movie film any perceptible point on the ice field (along a line coinciding with the pier's longitudinal axis), we determined the distance of this point to the cutting face of the pier.

We then measured this same distance on some other frame, which made it possible to determine the velocity of the ice field in some given time segment, since the number of movie frames taken per second was known and constant.

In this manner, we established the velocity v_η of the ice field prior to contact with the pier and also velocity v at any time moment t after contact with it.

The film of the movement of a small ice floe drifting through the same path as the ice field provided the possibility of establishing the surface current velocity v_0 in this river sector.

Length of ice field L could be measured based on the frames of the movie film as the sum of distances between a series of prominent points on the ice floe.

The area Ω of ice field was established with consideration of its form which was estimated on the basis of the floe's photographs or sketches.

In this manner, we determined all the values entering the working formulas.

For a refinement of the value of ice pressure, in recent years, the staff member at the Laboratory of Ice Thermics in the Siberian Branch of the USSR Academy of Sciences, V.K. Morgunov developed some supplements to the procedure, envisaging:

a) the application of a special long-distance camera permitting the obtainment, on one glass negative, of successive images of a moving object;

b) the application of a portable photogrammetric camers for determining the dimensions of an ice floe;

c) the application of the photographic transformation of pictures;

d) the dropping (on the ice floes) of markers with a metal tip which then became stuck on the ice and, showing up in the photographs and movie frames, would permit a more precise determination of the floe's velocity.

As the work experience has shown, these supplements, not changing the nature of the operation, undoubtedly promote a more accurate determination of the force of ice pressure.

The proposed equipment has been described below in more detail and also in reports [194,195].

4. New Equipment for Determining the Dimensions and Speed of Ice Field

Camera for Determining the Velocity of Ice Floe. The technique of photographing the process of ice pack's disruption during contact with a structure enabled us to expand considerably our concepts concerning the nature of interaction of an ice floe and a structure; it also permitted us to develop new methods corresponding to the actual pattern of the phenomenon, for the calculation of the ice loads. With the help of filming, we succeeded in determining the dimensions and velocity of ice fields, using the kinematic method of establishing the actual pressure of ice on structures.

However, there are obvious difficulties in processing the observational results. The dimensions of the movie frames are small, the material of the movie film is subject to some degree of shrinkage, which decreases to a certain extent the accuracy of establishing the ice loads.

To be sure, one should indicate that in a determination of the value of ice loads, we are unable to strive toward particularly high accuracy owing to the nature of the phenomenon per se. In individual years, the conditions of ice passage vary greatly, and the dimensions of floes, the rates of their motion, thickness and strength of ice are subjected to appreciable fluctuations. Nevertheless, we should strive toward a reduction in the scale of operation and to an increase in their accuracy within reasonable limits.

The camera designed by V.K. Morgunov has a long-distance objective (a focal distance equalling 300 mm in place of 35 mm in the movie camera) and it permits the obtainment of a single glass lens (18 X 24 cm) of successive images of a part of the ice floe, for example markers dropped onto an ice field drifting toward a structure (through short, equal time intervals).

This is achieved with the aid of a vertical slot with a light impervious shutter located in front of the photographic plate. During the observations, the shutter will move in such a way that the image of some given part of the ice floe is always located within the limits of the frame aperture.

The glass negative eliminates shrinkage and the large sizes of the image facilitate the processing of the photographs. When the photography is not conducted perpendicularly to the path of the ice floe's movement, the photographic transformation of the picture is possible, achieved by the standard techniques [194,196].

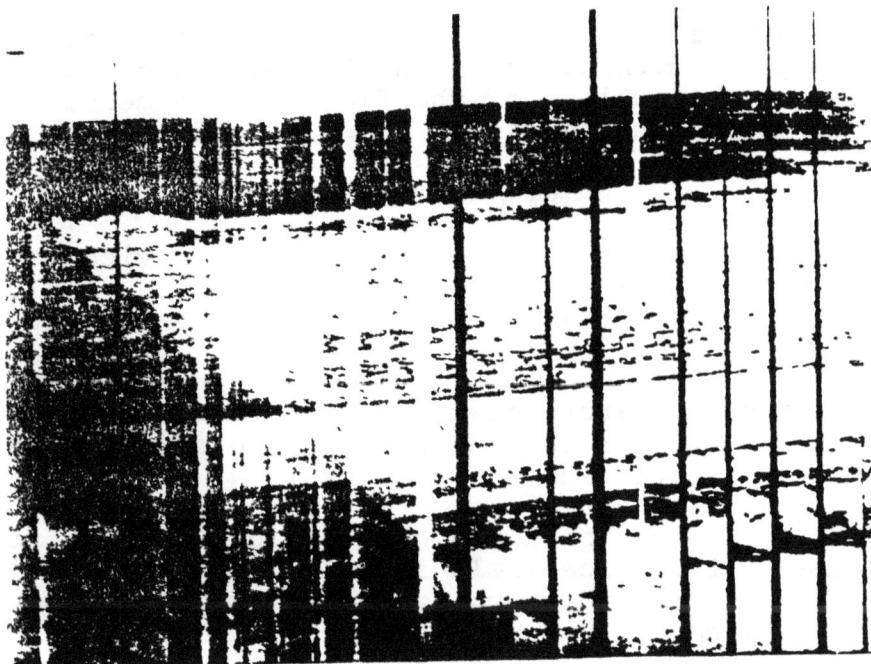

Fig. 64. Successive Positions of Markers on a
Moving Ice Field, Photographed with the Morgunov
Camera.

Figure 64 depicts the successive positions of two markers dropped onto a moving ice field. Knowing the spaces between the exposures, it is easy to determine the velocity of the ice field with the aid of a measuring magnifying glass, which permits the taking of readings with an accuracy of 0.03 mm. At a distance of 150 m from the device to the marker and a focal distance of 0.3 m, we have the opportunity of estimating the displacement of the ice floe by 20-30 mm.

Camera for Determining the Dimensions of an Ice Floe. In many working formulas, we include the dimensions of the floe and especially its area and thickness. It is not necessary to stress the difficulties developing in an estimation of the form and dimensions of ice floes approaching a structure. The portable photogrammetric camera proposed by Morgunov [194] is a rigid camera with a fixed inclination angle of the optical axis to the horizon. This feature permits utilizing one perspective grid for the photographic transformation of photographs and constructing a layout of the locality and a layout for the drifting ice floe.

Having a perspective grid in the photographs, we can construct a grid matching it in the layouts, and based on the cells, we can draw the outline of the floe. The dimension of a side of the grid square in the layout $d_{\text{лп}}$ can be found with the formula

$$d_{\text{лл}} = 1000 \ Hk \ 1/m_o \ \text{mm}, \tag{20.8}$$

where H = the excess of the center of the device's lens above the water level, m; $1/m_o$ = scale of layout; d_m = side of square for the locality, m; and $k = d_M/H$ = a constant value for each grid.

A second feature of the device is the application of a perspective grid for the layout by a photographic method during the process of printing the positive.

Fig. 65 is a photograph of an ice field approaching a structure. If the ice field is not placed in the viewing field of the device, it can be photographed by sections and we can obtain an outline of the ice floe after the matching of common points.

The camera permits a rapid and fairly precise determination of the sizes of passing ice fields. It is quite significant that as a result of the investigation, a document will remain, excluding the possibility of subjective estimations (evaluations)

Fig. 65. Application of Perspective Grid Permitting a
Determination of the Dimensions of a Drifting
Ice Floe.

Determination of ice thickness with the aid of photographing or filming contains, we might say, maximum difficulties.
A preliminary study of the ice thickness clarifies the picture
in a specific region, while ice fields are approaching the
structure from other sections of the rivers, at times located
at a distance of several tens of kilometers. In addition, the
study of the ice thickness stops 5-8 days prior to the ice passage, since this work becomes dangerous, and in connection with
this the determination of the ice thickness assumes special interest.

As was already mentioned above, for this purpose the
Laboratory of Ice Thermics at the Siberian Branch of the USSR
Academy of Sciences utilized a photographic and filming process
of the breakdown of an ice floe. We will describe the procedure
involved in conducting these operations [194].

In the upper area of a bridge support or of a spillway
(overflow) pier of a dam, a camera or movie camera is installed,
fixing the moments of the climbing of ice onto the sloped cutting

-257-

face or the processes of the "tilting" of the ice floes, also possible with vertical faces. In this connection, we should strive for a parallel condition of the optical axis of the camera for photographing the sloped face of the support, or a minimal deviation of 10-15°, since this reduces the error of observation.

In this connection, it is necessary that into the camera's viewing field, a part of the support would have entered, the facing joints of which should be applied to the drawing on a timely basis.

Running the photographic or movie film through a projector [194], we find the ice thickness of the screen and perform a conversion to natural conditions based on the formula

$$h = h_3 l /10nF,$$

(20.9)

where h = ice thickness in nature, cm; h_3 = ice thickness measured on the screen of the projector, mm; F = focal length, mm of camera l = distance from facing joint to center of equipment lens along the direction of its optical axis; and n = magnification of projector.

Let us note that it is most feasible to utilize cameras with wide-angle lenses for these purposes.

5. Brief Conclusions

1. The methods having existed until now of determining the ice pressure in nature required fairly expensive and complex equipment and therefore could not find broad application in actual practice.

2. The kinematic methods suggested by us for determining the pressure of ice in nature is very simple and permits us to find, without any special equipment, the actual ice pressure on the supports of various dimensions, form and material. For the production of observations, it is sufficient to have a small group of persons with 2-3 geodetic instruments during the period of the spring ice-out.

3. It is more feasible to utilize the filming of the process involved in breaking down the large floes by the structural supports, which will permit us to obtain all the data required for calculation.

4. A further refinement of the method (a more precise determination of the floes' sizes and velocity) can be accomplished with the aid of special equipment.

5. With a minimal outlay of time and resources, the kinematic method permits us to:

a) determine the actual resistance of ice to the penetration of piers into it;

b) to refine the effect of a pier's form in plane and profile; and

c) to explain the actual ice strength in the ice passage period based on individual regions in the USSR territory.

6.   The kinematic method is suggested only for determining the pressure developing during the cutting of a large floe to the point of its complete splitting. This pressure is the maximum of all possible ones and therefore it represents the maximal interest.

We have applied the kinematic procedure widely in finding the action of ice on the structural supports on the Ob', Tom', Yenisey and Angara Rivers from 1954-1961,(more than 40 determinations).

Chapters 21 and 22 are devoted to discussing the results of its application.

SIXTH PART

EXPERIENCE GAINED IN DETERMINING ACTUAL PRESSURE OF ICE ON
STRUCTURE BY USE OF KINEMATIC METHOD

Chapter XXI

EXPERIENCE IN UTILIZING KINEMATIC METHOD

1. Introductory Remarks

The question pertaining to the amount of actual ice
pressure on engineering structures assumes particularly great
significance at the present time when, in the eastern regions
of the country, construction of the largest hydroelectric
stations and new railroads and highways has been launched.

In connection with this, from 1954-1961, the Laboratory
of Ice Thermics at the Transport Power Engineering Institute
of the Siberian Branch of the USSR Academy of Sciences organized
field observations of the process of demolishing an ice pack
by piers under engineering structures (bridges, hydroelectric
power plants). In the utilization of the kinematic procedure
developed by us, we succeeded in determining the actual ice
pressure on piers with vertical and sloped cutting faces, pointed
and rounded in plane, with varying nature of their surface
(facing with hewn rocks, concrete, ferroconcrete).

It appears that such research can serve as a point of
departure for evaluating the various suggestions and standards
on determining the design loads and on their rectification. In
addition, they clarify the physical pattern of the phenomenon
and permit a more valid promulgation of systems for computing
the ice loads.

2. Scale and Results of Research Conducted

For estimating the maximum ice pressure on the piers, it
is necessary to observe a case when a large floe contacts more
or less directly with a pier and when at this time, the initial
cutting begins. For the springs in 1954-1955, we observed four
ice fields which satisfied the requirements imposed. In ad-

dition, we succeeded in using the observations conducted as early as 1933 on the cutting of a large ice field measuring 600 X 130 m, having struck against 5 bridge piers with sloped ice-cutting faces.

In 1956, we conducted observations of the functioning of bridge piers with a vertical cutting face, at which time we succeeded in using observations of the cutting of two ice fields. From 1957-1961, we conducted determinations of the actual ice pressure on supports of bridges and the buttresses of spillway dams with sloped and vertical faces. Of the numerous observations, four proved interesting for piers with sloped faces, and four for piers with vertical cutting faces. The results were processed in conformity with the procedure described in Chapter 20 and summarized in Table 56. The calculation details are clear from the examples listed below. Determinations No. 16-18 pertain to the ice action on structures of HES (hydroelectric stations) under construction.

As is evident from Table 56 pressure is closely associated with the nature of the ice-cutting face. While pressure on a pier with vertical face reached 62.6 tons, it did not exceed 45.8 tons on one with a sloped face. Thus, the sloped face greatly reduces the amount of pressure. One also notes appreciable fluctuations in pressure during the same spring. This can be explained by virtue of the natural heterogeneity of the ice pack, the presence of fissures, and the phenomenon of floes' splitting. For an illustration of the tendencies in the pressure drop, we list two typical graphs.

Figure 62 pertains to floe No. 6. The increased velocity in front of the bridge is caused by the form of the water surface (decay curve). In the period of the pier's cutting into the floe, pressure is slightly less, which is also reflected on the amount of acceleration. A velocity surge is noted and hence also a pressure surge, linked with the natural process of the breakdown and inevitable inaccuracies in the processing the movie film and the photographs.

Figure 66 reflects the curve for the velocity decrement for a floe with a vertical cutting face. Attention is drawn to the reduced scattering of the experimental points, which is explained in the following circumstances. The nature of the demolition of an ice field by crushing is not as violent as during bending. In addition, in this case, more precise equipment was utilized (see Chapter 20) and the observational errors were reduced.

Table 56

Results of Determining Actual Ice Pressure

No. of floe	Year Год	Dimensions of floe			v_0, m/sec	v_Λ, m/sec	v_1, m/sec	t, sec	H, m tons	Specifications of pier (support)
		Ω, thous. of m²	L, м	h, м						
1	1933	15,6	130	0,98	1,80	2,16	0,20	107	43,8	$b_0 = 4{,}4$ м
2	1954	2,5	48	0,50	1,30	1,30	1,15	2	17,5	$2\alpha_1 = 19°$
3	1955	24,0	175	0,80	2,50	2,80	2,40	4	46,4	$r = 1{,}0$ м
										$\beta = 57{,}5°$
4	1955	7,8	102	0,65	2,00	2,20	1,54	7,5	45,3	$2\alpha = 46{,}5°$
5	1955	1,6	46	0,90	2,50	2,76	1,80	2	21,2	
6	1959	9,7	189	0,50	—	0,70	0,60	10	11,0	$b_0 = 4{,}0$ м
										$r = 2{,}0$ м
7	1959	21,8	200	0,60	—	0,70	0,64	10	8,5	$\beta = 45°$
8	1959	0,74	33	0,85	—	1,96	1,47	2	15,2	$2\alpha = 80°$
9	1959	1,05	50	0,75	—	2,60	1,56	4	11,5	
10	1956	3,6	60	0,50	1,00	1,38	0,62	2	62,6	$b_0 = 7{,}7$ м
11	1956	44,0	200	0,40	1,00	1,33	1,17	4	55,0	$2\alpha = 115°$
										$r = 0$
										$\beta = 90°$
12	1960	3,6	75	0,70	—	1,63	1,15	4	28,3	$b_0 = 3{,}4$ м
										$r = 0{,}6$ м
13	1960	2,88	54	0,70	—	1,85	1,61	2	23,0	$2\alpha_1 = 92°$
14	1960	0,3	28	0,70	—	1,72	0,53	1	22,0	$2\alpha = 88°$
15	1960	0,9	42	0,70	—	2,08	1,82	1	16,0	$\beta = 90°$
16	1960	0,8	—	0,80	—	1,80	0,90	4	14,0	$b_0 = 12$ м
17	1960	1,7	—	0,80	—	1,90	1,65	2	15,0	$\beta = 90°$
										$2\alpha = 180°$
18	1961	26,2	—	0,90	—	1,23	0,63	60	22,5	$b_0 = 12$ м
										$\beta = 90°$
										$2\alpha = 180°$

Let us point out that the simultaneous observation of floe from the side (from an adjoining pier) and from (from the given pier) permits us to explain the disruption in the regularity of the curve for velocity's dependence on time of cutting, as e.g. is clear from Fig. 62.

For illustrating the methods in computing the actual pressure, we have presented two examples below for floes No. 1 and 4.

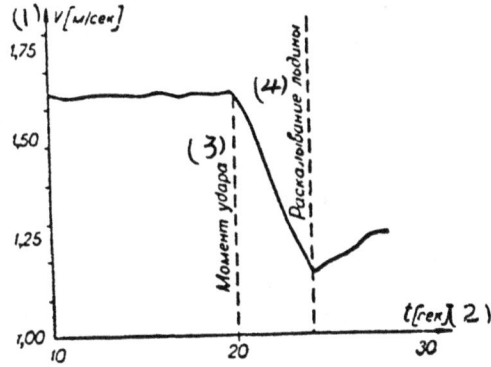

Fig. 66. Variation in Floe Velocity After Contact with
Vertical Face of Pier. Key: 1) v[m/sec]; 2) t[sec];
3) time of impact; and 4) splitting of floe.

3. Determination of Ice Pressure Under Natural Conditions

Example 1 (for floe No. 4)

Initial Data. On 25 April 1955, an ice field with length
L - 102 m and a rounded form was cut by a bridge pier with width
b_o = 4.4 m. The floe moved at velocity v = 2.2 m/sec. The
filming of the floe breakdown process established its thickness
at 0.6-0.7 m. The floe's area was assumed at Ω = 0.78 L^2=7800 m^2

The film recorded the drop in floe's velocity after contact with the pier. The nature of the drop is reflected in the
following figures:

t_1 sec :   0   1.2   3.3   5.0   7.5

v_1 m/sec :   2.2 2.0   1.8   1.68   1.54

The ice was broken by bending.

System of calculation. The floe's mass:

$$m = \frac{\gamma \Omega h}{g} \quad \frac{0.92 \cdot 7800 \cdot 0.65}{9.8} = 477 \quad \frac{m/sec^2}{m}$$

-263-

The floe's acceleration at various time periods proved
to be:

for the period 1.2 - 3.3 sec, j = 2-1.8/3.3-1.2 = 0.095 m/sec^2

1.2 - 5.0 sec, j = 0.084 m/sec^2 and

1.2 - 7.5 sec, j = 0.073 m/sec^2.

The maximum pressure (for period 1.2 - 3.3 sec) comprises:

H = m j = 477·0.095 - 45.3 tons,

for period 1.2 - 7.5 sec respectively

h = 477 X 0.073 = 34.8 tons.

Let us note that in effect, the same results could have been obtained by utilizing the procedure developed by us for the first,
case. Let us demonstrate this in the actual data of the same
example.

Another Calculation Model. The coefficient of total
resistance to the floe's motion (based on the author's empirical
formula substantiated above)

$$f = \frac{\sqrt{h}}{100\sqrt{L}} = \frac{0,65}{100\sqrt{102}} = 0,00081.$$

The calculation is conducted based on the formula

$$v_t = v_0 - \sqrt{\frac{c}{f\Omega}} \, \text{th} \frac{(t_1 - t_0)\sqrt{cf\Omega}}{M}$$

with the use of the curve shown in Fig. 63.

For time moment t_i = 3.3 sec, we have:

$$A_1 = (v_0 - v_t)\sqrt{f\Omega} = (2 - 1,8)\sqrt{0,00081 \cdot 7800} = 0,50;$$

$$B_1 = \frac{(t_t - t_0)\sqrt{f\Omega}}{M} = \frac{(3,3 - 1,2) \, 2 \cdot 51}{477} = 0,011.$$

Based on the curve (Fig. 63), we find c = 45 tons and

$$H = c + f\Omega(v_{\Pi} - v_o)^2 = 45 + 0.00081 \times 7800 \ (2.2-2)^2 = 45.3 \text{ tons.}$$

Analogously, for moment $t_1 = 7.5$ sec, we have:

$$A_2 = (v_o-v_i) \ \sqrt{f\Omega} = (2 - 1.54) \times 2.51 = 1.15;$$

$$B_2 = \frac{(t_i-t_o)\sqrt{f\Omega}}{m} = \frac{(7.5 - 1.2) \times 2.51}{} = 0.033.$$

According to the curve

$$c = 35 \text{ m and } H = 35+0.00081 \times 7800 \ (2.2-2)^2 = 35.2 \text{ tons.}$$

$$(H_{max} = 45.3 \text{ tons}).$$

Example 2 (for floe No. 1)

On 23 April 1933, on a large river in Siberia, we observed the following event. A floe measuring 600 X 130 m moved obliquely, striking against one of the piers, turned and was then cut by five bridge piers without splitting. Concentration of ice movement was slight and the confining effect of other floes could be disregarded.

Current velocity was 1.8 m/sec; velocity at end of floe's cutting comprised around 0.2 m/sec. Based on 147 measurements, ice thickness proved to average 0.98 m.

The pier was solid, 4 m in width, with angle of inclination to the horizon $\beta = 57°30'$ and with rounded outline of cutting face.

Data are very scant and therefore it is necessary to use the first method. Let us describe the process of solution.

If the floe under consideration had not experienced the restraining effect of other ice fields, its velocity would have equalled

$$v_{\Pi} = (1.1 - 1.2) \ v_o,$$

or an average

$$v_{\Pi} - 1.2 \ v_o = 2.16 \text{ m/sec.}$$

As was shown in Chapter 19, the decrease in the floe's velocity in the initial period of the pier's penetration into the ice occurs essentially in accordance with a linear law, in connection with which we can assume the cutting rate to equal on an average

$$v_{av(cp)} = (2.16 + 0.2)\ 0.5 = 1.18 \text{ m/sec.}$$

Since the length of the floe's cut sector constituted 130 m, the cutting time can be assumed to equal

$$T = 130 : 1.18 = 110 \text{ sec.}$$

In Eq. (20.3), by t we connote the time having elapsed from the beginning of the third stage, i.e. from the moment when the floe's velocity declined to the value of 1.8 m/sec. Assuming the linear law of decline in floe's velocity, we find by interpolation x = 24 m.

Consequently the time of cutting the floe in the third stage will be determined as follows:

$$t = T - x/v_{av} = 90 \text{ sec.}$$

The floe's mass:

$$m = \gamma \Omega h/g = 1440\ \frac{\text{m.sec}^2}{\text{m}}\ .$$

Let us determine the general coefficient of flow's resistance based on an approximate formula (since L < 300 m):

$$f = \frac{\sqrt{h}}{100\sqrt{L}} = 0.00088\ \frac{\text{m.sec}^2}{\text{m}^4}$$

$$b = f\Omega 13.7\ \frac{\text{m.sec}^2}{\text{m}^2}$$

Let us find in addition

$$A = (1.8 - 0.2)\ \sqrt{13.7} = 5.90$$

$$B = (90\text{X}\ \sqrt{13.7}) : 1440 = 0.23$$

Based on the curve in Fig. 63, we find c = 42 tons and

$$P = c + f\Omega(v_\Lambda - v_o)^2 = 43.8 \text{ tons.}$$

As we shall see, even the extremely limited data permitted us to estimate the order of the ice pressure's magnitude However, there can be no doubt that more complete and reliable data can be obtained with the availability of a complete decay curve of the floe's velocity when an ice field is being cut through by a pier.

Chapter XXII

COMPARISON OF ACTUAL PRESSURE OF ICE WITH CALCULATED PRESSURE

1. Introductory Remarks.

In connection with the broad scale of construction in the USSR in recent years, a number of proposals have appeared on the calculation of ice loads on structures. Let us note that since 1946 in the USSR, even the Technical Specifications and Standards on determining the ice loads have changed 2 - 3 times. The Union-wide standard, GOST 3440-46, that had been in effect since 1946, was superseded on 1 April 1960 by the new standards contained in SN 76-59. The USSR Ministry of Railways and the USSR Ministry of Transport Construction replaced in 1947 standards (TUPM-47) in 1956 by new ones (TUPM-56) which in respect to ice loads became obsolete after the publication of SN 76-59.

Such a frequent change in the standards is explained by the explicit nonconformity of certain of them with the physical pattern of the phenomenon. Technical Specifications SN 76-59 now in effect are also in need of corrections and improvements, as we have indicated in Chapter 8.

In connection with this, it is necessary to compare the derived values of the actual ice pressure with the standards and also with the suggestions made by a number of researchers. The comparison was conducted with the requirements of GOST 3440-46, TUPM-47, TUPM-56, SN 76-59, and also with the recommendations made by B.V. Zylev and the author.

2. Mathematical Relationships

The working formulas for determining the maximal ice pressure on structural supports are presented in Table 57.

Notations adopted: b_o = width of pier in plane; 2α = angle of pier's tapering; s_o, m = coefficient of pier's form; r = radius of cutting face's rounding; R_p, R_M, R_c = ultimate crushing, bending and shearing strength of ice; β = inclination angle of cutting face to the horizon; and h = ice thickness.

Table 57

Summary of Working Formulas

Date of proposal	Author	Type of cutting face			
		Vertical	R_p, т/м²	Sloped	$R_н$, т/м²
1946	GOST 3440—46	$P = R_p b_0 h$	100—200	$H = P\sin^2\beta$	
1947	TURI —47	$P = kvh\sqrt{\Omega}$ with cutting-face without " face	50—100 35—75	$H = P\sin^2\beta$	
1956	TURI —56	$P = k_n b_0 h$ $P = ch\,(ar+bB)$	50—100	$H = 1,5 P_c$	
1960	SN 76—59	$P = m R_p b_0 h$	45—150	$H = R_н h^2 \lg\beta$	31—105
1950	B.V. Zylev	—	—	$H = \dfrac{R_н h^2 \lg\beta}{A}$	150
1951— 1960	K. N. Korznavin	$P = K_p b_0 h$ $K_p = \dfrac{2,5\,mkR_p}{\sqrt[3]{v}}$	40—60	$H = R_н h^2 \lg\beta\,\dfrac{b_0}{h}\,S_0$	40—60

3. Examples of Calculation

Floe No. 4. A floe with an area of 7800 m², thickness of 0.65 m on 25 April 1955 moving at 2.2 m/sec struck against a solid bridge pier. The pier's dimensions (see Fig.47) are typified by the following data: width b_0 = 4.4 m, tapering angle $2\alpha_1$ = 19°, rounding radius of cutting face r = 1 m, and angle β of its inclination to the horizontal = 57.5°.

Then the calculated tapering angle:

$$2\alpha = 2\alpha_1 + \frac{4r(40° - \alpha_1°)}{b_0} = 46°30'.$$

The results of calculation:

according to GOST 3440-46

$$H = R_p b_0 h \sin^2\beta = 100 \cdot 4.4 \cdot 0.65 \cdot 0.84^2 = 206\ t$$

according to TUPM-47

$$H = k_{\text{Л}} \, b_o h \sin^2 \beta = 50 \cdot 4.4 \cdot 0.65 \cdot 0.84^2 = 103 \text{ t;}$$

according to TUPM-56

$$H = 1.5 \, \frac{R_p b_o hc}{D} = 1.5 \, \frac{50 \cdot 4.4 \cdot 0.65 \cdot 2.00}{32.8} = 13.2 \text{ t;}$$

according to SN 76-59

$$H = R_{\text{И}} \, h^2 \tan \beta = 0.7 \cdot 150 \cdot 0.65^2 \cdot 1.57 = 70 \text{ t;}$$

according to B.V. Zylev

$$H = \frac{R_{\text{И}} \, h^2 \tan \beta}{A} = \frac{150 \cdot 0.65^2 \cdot 1.57}{1.11} = 89.7 \text{ t}$$

at

$$A = \frac{1.1}{1.2\eta + 0.56 \dfrac{h \, tg \, \beta}{\sqrt[4]{\bar{D}}}} = 1.11;$$

$$D = \frac{Eh^3}{12(1 - \mu^2)} = \frac{300000 \cdot 0.65^3}{(1 - 0.4^2) \cdot 12} = 8060; \quad \eta = 0.96;$$

according to K.N. Korzhavin

$$H = R_{\text{И}} \, h^2 \tan \beta \, (b_o/h) \, S_o = 50 \cdot 0.65^2 \cdot 1.47 \, \frac{4.4}{0.65} \cdot 0.23 = 51.0 \text{ t}$$

(at coefficient of form S_o = 0.23, according to Table 34, Chapter 8, 6).

As was shown above, the actual pressure for floe No. 4 was found to be 45.3 tons. The combined results of the calculations are illustrated by the following data:

Actual pressure of 45.3 t -- 100%
According to GOST 3440-46, 206.0 t -- 454%
According to TUPM-47, 103 t -- 227%
According to TUPM-56, 13.2 t -- 29%
According to SN 76-59, 70.0 t -- 154%
According to B.V. Zylev, 89.7 t -- 198%
According to K.N. Korzhavin, 51.0t - 112%

Floe No. 11. An ice field with area Ω = 44,000 m^2, thickness of 0.4 m, moving on 17 April 1956 with a speed of 1.33 m/sec, struck against a solid bridge pier. Actual pressure was 55.0 tons.

Dimensions of support:

b_o = 7.7 m, 2α = 115°, β = 90°.

Results of calculations: according to GOST 3440-46

$$P = R_p b_o h = 100 \cdot 7.7 \cdot 0.4 = 308 \text{ t;}$$

after TUPM-47

$$P = R_p \bar{b}_o h = 50 \cdot 7.7 \cdot 0.4 = 154 \text{ t;}$$

after TUPM-56

$$P = \sigma h(ar + bB) = 100 \cdot 0.4 \cdot 5.45 = 218 \text{ t;}$$

after SN 76-59

$$P = m R_p b_o h = 0.79 \cdot 45 \cdot 7.7 \cdot 0.4 = 109 \text{ t;}$$

after K.N. Korzhavin

$$H = K_p \, b_o h = 36{,}3 \cdot 7{,}7 \cdot 0{,}4 = 112 \text{ t;}$$

$$K_p = \frac{2{,}5 R_p m k}{\sqrt[3]{\bar{v}}} = \frac{2{,}5 \cdot 50 \cdot 0{,}80 \cdot 0{,}40}{\sqrt[3]{1{,}33}} = 36{,}3 \text{ t/m}^2.$$

As we shall observe, the recommendations of the various standards show a higher pressure than that which occurred. This is natural because as a rule the design pressure should be higher than that which is observed.

4. Results of Comparing the Actual Pressure of Ice with the Computed Pressure

As we see from Table 56, all the observations were conducted at 5 types of piers, having various dimensions and form. In connection with this, it is sufficient to compare the maximal pressure in each observation year with the design pressure at maximal ice thickness. The calculations are given in Table 58.

5. Brief Conclusions

1. The suggestions of a number of authors for finding the ice loads on the supports of bridges and hydraulic engineering structures need to be reviewed.

2. In a review of the standards, one should rely on field observations and theoretical developments, corresponding to the physical nature of the phenomenon. In view of the considerable heterogeneity of ice in the ice-out period and the absence of a linear relationship between the stresses and deformations, the application of the theory of elasticity obviously is unable to yield satisfactory results.

Table 58

Comparison of Actual Ice Pressure with Design Pressure

Data	Unit of measurement	Year of observation					
		1933	1954	1955	1959	1956	1960
Number of floes	m²	1	2	3—5	6—9	10—11	12—15
Area of ice field. . . .	thous.	15,6	2,5	24,0	0,7	3,6	3,6
Thickness of ice	m	0,98	0,50	0,90	0,85	0,50	0,70
Dimensions & form of pier, width	»	4,4	4,4	4,4	4,0	7,7	3,4
r — radius of rounding .	»	1,0	1,0	1,0	2,0	0,0	0,6
2α — tapering angle .	degr.	46	46	46	80	115	83
β — inclination angle .	»	57	57	57	45	90	90
Actual pressure	tons	44	18	47	15	63	28
Calculated pressure per GOST — 3440—46 . .	»	304	156	280	170	178	238
per TUMP —47 . . .	»	152	78	140	85	192	119
per TUMP —56 . . .	»	27	9	24	23	137	62
per SN —76—59 . . .	»	95	26	30	46	137	79
per B.V. Zylev	»	203	55	165	96	—	—
per K. N. Korzhavin . .	»	78	40	72	27	108	70

Remark. In 1954 and 1960, the ice strength was less than usual.

3. It is extremely desirable to determine the actual ice pressure on structures by undertaking the appropriate field observations and comparing the data derived with the suggested calculation techniques.

4. For this purpose, the kinematic method is most convenient; at a minimal outlay of resources, it permits the accumulation of the necessary information concerning the actual ice pressure.

5. Experience in using this method has shown that the recommendations of GOST 3440-46 and TUPM-47 led to an exaggeration of the design value by 5-9 times; for the case of piers with a sloped cutting face, the TUPM-56 recommendations give very under-estimated results, about half as great as those observed; calculations with Zylev's method lead to results close to the TUPM-47 requirements; and the SN76-59 recommendations for the case of piers with an inclined cutting face yield elevated design forces in part as a result of the unjustifiably high values for the ultimate bending strength of ice.

It is curious that based on the recommendations of SN 76-59, the calculated pressure of ice is higher in some cases than the ice loads which actually occurred by 1.4 -2.4 times, while in other cases, the discrepancy is even more (up to 3.5 times). In our view, this can be explained by the failure to allow for the undisputed effect of pier width on the value of the ice loads. Based on our proposals, the calculation of the ice loads yields results greater by 1.6 - 1.8 times than those which occurred, and moreover, this ratio is fairly constant. Just in one case (floe No. 2), the design (calculated) value is 2.4 times higher than the actual one. From our viewpoint, this is indicative of the correctness of the structure in the suggested formulas. Taking into account that the observations were conducted for an ice pack of average or partly even of weak strength (floe No. 2, 12-15), we consider it possible to utilize the formulas proposed.

6. A further processing of the questions should be conducted in the direction of finding the actual value of ice pressure on structural supports, which will permit the introduction of corrections into the proposed standards, the provision of a regional division of the USSR by amount of ice pressure, and the recommendation of the most feasible designs of structures absorbing the ice pressure.

7. It is necessary to coordinate the efforts of the scientific-research organizations, working in the area of determining the ice pressure on structures, and to organize a broader exchange of the research findings.

SEVENTH PART

PASSAGE OF ICE AND SIZE OF ICE-ADMITTING OPENINGS

Chapter XXIII

FEATURES OF ICE PASSAGE IN LINE OF STRUCTURE

1. Introductory Remarks

In the construction of the HES on the rivers with considerable ice-passage, we often find the holding of ice in the
upstream water, excluding its discharge into the downstream
water.

However, at this time there is a shortening of the navigation period and possibly the formation of jammings in the zone
of backwater's tapering-out. In addition, we should take into
account the distribution of water discharges between the dam
and the station, and the value and velocities of current in the
upstream water. At an increase in velocities (up to 0.4 - 0.5
m/sec) it is possible to have the opening of the reservoir,
coinciding in time with the maximal discharge requiring the
release of water [65].

Finally, during the construction of the HES, for several
years, it is necessary to admit ice into the site of the structure.

In connection with this, it is useful to review:

a) the methods of taking into account the ice passage con
ditions during the installation of a unit for the structure in
plane;

b) the operating features of the piers of the spillway
dams;

c) dynamic ice pressure during discharge into the downstream water;

d) measures facilitating the passage of ice through the structures.

The present chapter is devoted to a discussion of these questions.

2. Allowance for Conditions of Ice Passage with Utilization of System of Structures' General Setup

An efficient solution of this problem is possible only under actual given conditions with a detailed consideration of local features, technical-economic concepts and conditions of the performance of a range of construction operations. Therefore, we will limit ourselves only to the presentation of general policies on this question with consideration of literature data [43, 148, 180 etc.] and personal observations.

1. In choosing the location of the dam of a HES or bridge on the rivers with severe and heavy ice passage, it is desirable to have above the structure a long river sector, as straight as possible.

2. In placing the structures in the bend of a river, it is desirable to have spillway spans of the dam (utilized in the construction period for the passage of ice) at the concave shore, toward which the main mass of ice is usually directed. In dividing the bridge opening into spans, one should avoid small spans at the convex shore since as a result of the reduced velocities and depths, a frequent stoppage of ice fields is possible here.

3. The presence of an island or submerged shoal above the low-pressure structure is undesirable, since this creates an irregular distribution in the ice masses along the spillway front and the sloping of the nappes. Let us note that for the operation of the arches being erected for cutting off the trough of the HES, the presence of a shoal located above it is also desirable. In the ice-passage period, on the shoal, accumulations of ice develop, reliably protecting the structures from the further effect of ice.

4. One should not locate the engineering structures (bridges, dams, water intakes) below the river sectors with the frequent formation of ice jams. After the breakup of the jam, it is possible to have the movement of large ice accumulations with high velocity, which does not exclude serious damage to the structure.

Of course, this does not pertain to the large dams, the pressure from which varies the ice conditions of a river appreciably and excludes the possibility of the further formation of jams in the given site.

5. It is useful to locate the water-collecting structures on the side away from the direction of main ice masses and in calm water. In this case, a rapid ice jam reduces the difficulties from intra-water ice to a minimum.

6. In developing a plan for the performance of construction tasks, one should set up the sequence for erecting the various structural elements with a careful consideration of their function in the ice-passage period. The passage of a strong ice-out in 1960 at the Mamakanskaya HES [148] would have been difficult if all of the upper piers had been erected simultaneously and the opening of the span would accordingly have been narrower.

7. The ice-protecting panels or walls (if they are built) should be planned with consideration of the requirement for a smooth direction of flow (at an angle of 65°-90°) to the water-admitting openings.

8. In connection with the complexity of solving many problems, it is desirable to practice more broadly the preliminary studies on hydraulic models. Unfortunately, a technique for modelling ice has not yet been developed and the question has not yet been solved concerning the simulation of ice passage conditions not only with consideration of the extents and rate of floes' movement, but also particularly of the mechanical properties of ice. The practice of utilization of paraffin tablets or natural ice plates (even specially processed) for this purpose naturally does not guarantee the observance of the necessary conditions of similarity.

3. Certain Operating Features of Individual Piers of Overflow Weirs

The methods discussed in the third part for determining the ice pressure on structural supports pertain entirely to the individual piers located on the crest of the overflow dams. However, there are certain features of such structures which it is necessary to elucidate.

The ice regime of a pressure pier differs appreciably from the standard one in connection with the abrupt increase in depths, decrease in current velocities, variation in temperature regime of basin and increase in role of the wind.

As is known, the clearing of large reservoirs of ice occurs on an average of 10-12 days later as compared with a river in a natural state while in individual years, ice passage does not occur and the ice thaws in place.

The strength of ice decreases perceptibly as a result of the late opening of the river and the erosion of ice by the warmer water. D.N. Bibikov and N.N. Petrunichev [66] recommend determining the drop in compressive and shearing ice strength in the pre-ice passage period based on the formulas:

$$R'_{cx} = R_{cx} - az,$$
$$R'_{cp} = R_{cp} - bz$$

Fig. 67. Passage of Ice Through an Overflow Dam.

at $a = 1.5$ Kg/cm^2 in days; $R_{compres} = 35$ Kg/cm^2; $b = 0.3$ Kg/cm^2 per day; $R_{sh} = 10$ Kg/cm^2, and $z =$ number of days.

In this manner, we can consider that in a delay of the opening by 10 days, the ice strength will decrease significantly and will comprise only 0.70-0.60 of the ice strength up to the beginning of spring ice-out in the uncontrolled river sectors.

Detailed studies of the physico-mechanical properties of an ice pack conducted by the Laboratory of Ice Thermics at the Siberian Branch of the USSR Academy of Sciences in the large reservoirs of Siberia confirm this estimation.

Finally, we should point out that the delay in the ice passage is accompanied by a significant decrease in the thickness of the ice pack up to the time of opening (by 20 - 40% as compared to the maximum values recorded during the winter).

In the given case, the conditions of breakdown of the ice pack are complicated by the presence of a decay curve, developing in front of the spillway (see Fig. 5). During the approach to the spillway crest, the ice pack breaks up, thereby facilitating the work of the piers. Observations by Prof. V.T. Bovin [153] of the passage of ice through the dam at the Volkhovskaya Hydro-electric Station indicated that during the approach of an ice field, a crack is formed, oriented along the dam's axis and located at a distance of 3-4 H. In its turn, the broken-off part of the floe divided into two pieces above the overflow crest, while during the descent along the spillway surface of the dam, it cracked further into 3-4 parts, each of which was not larger than 3-5 m.

The stopping of large fields at the dam piers is possible but it should be kept in mind that in the practice of operating the dams, cases of jamming on the crest have not been recorded.

The photographing and filming of the processes of the ice's demolition indicate a more frequent splitting of floes and their subsequent breaking along the decay curve. One should also note the effect of local backwater in front of the piers (facilitating the demolition of the ice and undoubtedly reducing the value of the ice loads (Fig. 67).

In this manner, the piers of the spillway dams function under somewhat better conditions as compared with the bridge piers The most likely result in this case is the breakdown of floes from splitting (at vertical cutting face) or even from shearing in case of a sloped ice apron. Disruption from bending is less likely in connection with the greater than usual development of the spring thawing processes accompanied by an appreciable reduction in the ultimate shearing strength of ice along the crystals' axis.

Serious attention should be directed to imparting to the piers' cutting part of forms which are streamlined and tapered in plane.

Fig. 68. Sliding of Ice into the Structure's Downstream Water.

In most cases, the use of an inclined cutting face is
superfluous since the breakdown of the ice cover even without
a sloped ice apron is greatly facilitated by the presence of the
decay curve. Let us point out that the application of piers
with an excessively steep ice apron (1:17) sometimes being
practiced of course does not make sense.

4. Dynamic Pressure of Ice During Discharge Into Downstream Water

In the dumping of ice into the tailwater, it is possible
to have a significant dynamic effect even of floes which are
small in dimensions in plane but at times still fairly strong.
The ice pack, breaking up the decay curve, acquires high velocities of the order of 6-10 m/sec and more and individual floes
have a considerable reserve of kinetic energy (Fig. 68).

It is evident that the amount of developing pressure can
not be greater than the ultimate ice crushing strength, i.e. of
the order of 80-200 t/m^2 but apparently (with allowance for
impacts by a corner, by an uneven edge) even less.

However, also of considerable importance is the fact that
the impacts of floes in the identical parts of the structures

occur for a prolonged time, during several days of spring ice passage, from year to year.

The recurrent impacts cause local damages, i.e. the chipping and abrasion of the concrete surface, the shearing off of, or damage to, the projecting built-on parts. Interesting data have been reported [148] concerning the passage of the spring ice in 1960 at the site of the Mamakanskaya HES which was then under construction.

During the ice-out period, the upper edges of the pressure face of the sections in the ice passage season were chipped and rounded off; all of the offsets, projecting for 15 cm, were worn down flush with the basic concrete, which was also abraded for a depth up to 3-7 cm. The ribs of the lower piers were rounded and the piers acquired a streamlined form; right angles were gouged for a depth of 50-60 cm; the built-up parts were torn away along with pieces of concrete from a depth of 50-70 cm.

As we shall observe, as a result of just one (but severe) ice passage (ice thickness 0.8 - 1.4 m, thickness of isolated floes up to 3 m, area up to 200-400 m^2, rate of their movement within the structural site, 4.5-6 m/sec), localized but appreciable damages were caused. The situation was complicated by the hydraulic features of the flow (driven-off hydraulic jump, strong eddies).

To protect the structure from damages, during the dumping of ice into the tailwater, it is first of all necessary:

a) to develop a hydraulic flow regime assuring the smooth junction of the tailwater with the descending stream.

b) to provide the required depth of overflowing layer with an increased depth reserve in the lower part of the spillway dam of practical profile (up to two ice thicknesses).

c) to impart to the lateral faces of the spillway piers a streamlined smooth surface, without projections, and with a smooth rounding of edges.

d) to utilize concrete of increased strength for protection from the abrasive action of the ice.

5. Nature of Measures for Facilitating Passage of
Ice Through Structures

1. Regulation of river discharges in the HES cascade in the ice-out period, permitting a significant effect on its nature

2. The longest possible holding of the ice in the head water, permitting one to count on a decrease in the ice pack's thickness and strength.

3. Provision of the uniform operation of the openings in the spillway dams, preventing an accumulation of ice at certain of them.

4. Provision of hydraulic conditions of flow movement through the spillway with a smooth connection of the tailwater with the descending stream.

5. Use of icebreakers for reducing the dimensions of floes and for extending the navigational season.

6. The shaping of the spillway dam piers with smooth, tapered configurations, permitting them to split the approaching ice fields quite efficiently.

7. The careful development of a plan for the performance of a series of construction tasks with allowance for the necessity of ice passage during the period of erecting the structure.

8. The application of the necessary measures of an operational nature (clearing areas of ice, combatting the freezing over of ice jams, installation of temporary floating barriers, etc.).

A detailed discussion of the operating methods has been given in the reports written by A.N. Komarovskiy [1,43,154], D.N. Bibikov and N.N. Petrunichev [66], F.I. Bydin, A.M. Yestifeyva [155] and a number of other authors.

In conclusion, let us point out the method (recommended by A.F. Vasil'yev [180]) for filling the annual control shell of the reservoir (developed by the beginning of spring flood stage to the dead storage level) according to the following pattern.

First Stage. Leaving the reservoir level at the dead storage level, in proportion to the increase in extent of flow in the reservoir, we increase the output of the HES by the successive activation of all turbines. This will lead to an increase in velocities under the ice and will provide a more rapid thawing of it. A considerable discharge of water promotes freeing the ice from the adjoining tailwater sector.

Second Stage. After the volume of flow exceeds the total output through the turbines, filling of the reservoir begins without empty discharges for the purpose of accelerating the rise in level.

In the third stage, one should start the empty discharges with the plan of filling the reservoir by the time of the drop in extent of flow prior to full-scale discharge (output) through the HES turbines.

Chapter XXIV

DETERMINATION OF DIMENSIONS OF ICE-DISCHARGING OPENINGS

1. Introductory Remarks

Apart from the technical-economic considerations, the laying out the division of a structure into spans, one should take into account the necessity for the unjammed passage of ice. Analyzing the data on the damages and wear to the structures, caused by the action of the ice pack, it can be established that in most cases, they constituted the result of the formation of jams, having led to a rise in water level, the toppling of the bridge span structures or to the demolition of the piers.

If a part of an opening is covered by small spans, the opening can fail to function during the most responsible period of the spring ice debacle. Thus, at one of the large bridges in Siberia, spans with a length of 18 m during ice-out became systematically plugged with ice and did not take part in the ice-passing operation of the opening, while the spans of greater width admitted the ice successfully.

It is natural that the greatest difficulties in the passage of ice occur with the large floes of considerable strength, i.e. in the initial ice passage periods. As is known, during this time, it is quite possible to have the stoppage of the ice fields at the piers of bridges and structures; however, this does not represent any special hazard. After the thrusts, we usually have a rise in water, an increase in current velocity (and hence also in the force of floes' attraction by the stream), in connection with which the stopped ice fields are caused to move and become cut through by the piers. However, during the period of full-scale ice passage, the stopping of the floes can lead to the formation of an ice jam and to an intolerable rise in the water level.

As we shall see, the minimal width of a structure's span should not be less than a specific value, and moreover one should take into consideration:

a) necessity for the jam-free passage of ice;

b) concepts of a technical-economic nature; and

c) requirements ensuing from the conditions of shipping and timber floating (for the bridges).

2. Allocation of Width of Ice-Discharging Openings

In planning the structures on rivers with intensive and heavy ice passage, it is necessary to choose a width of the ice-admitting openings. This question has its own background. In the past, many authoritative hydraulic engineers have considered that on the rivers with a significant ice passage, the only dependable solution is the installation of a "blind" spillway, i.e. one not divided by piers or having gates on the crest.

However, experience in operating the dams has shown the exaggeration of the dangers which had existed, in connection with which the spillway dams with dividing piers and gates on the crest became quite popular. No cases are on record of jams on the dam crests and the temporary ice accumulations which did occur were soon broken up and dumped into the tailwater.

For the selection of the required width of the ice-carrying sector, the following methods were utilized [43,150,60].

A. Allowance for Features of a River Under Natural Conditions

Based on the configuration of a riverbed under natural conditions, we can clarify the dimensions of the narrower flow sectors, providing the unjammed passage of ice. An analysis of the data available for a number of USSR rivers permitted us to suggest the following formula for finding the width B_Λ of the ice-carrying sector:

$$B_\Lambda = b_1/b_2 \ B_\Pi = (0.30 - 0.45) \ B_\Pi , \quad (24.1)$$

where B_Π = width of river based on surface of supported water above the dam; b_1 = width of narrow river sector ahead of dam structure; and b_2 = width of river in a sector located directly above the narrow sector.

It is natural that this method can be regarded only as approximate, especially under the conditions of high dams markedly altering the river conditions.

B. Allowance for Work Experience with Completed Structures

A detailed study of the conditions of ice passage through the completed structures at a number of large hydraulic stations in the USSR permitted S.S. Agalakov [207], A.F. Vasil'yev [180] and other authors to formulate the basic conditions providing for the unjammed passage of ice.

For this, it was necessary that the extent of crowding the riverbed with arches would not be excessive, and that the length of the ice-carrying sector and the dimensions of individual spans would be adequate, just as the depths within the limits of the structure.

In Table 59, we have listed certain data on the sizes of ice-admitting structures on specific rivers in the USSR and abroad.

An analysis of the data listed in the table permits us to conclude that for most of the modern, successfully operating dams on rivers with a significant ice passage, the width of the ice-carrying sector comprises 0.45 -0.70 of the river's width in the supported canal between the dams; the individual spans have a size of 15-25 m.

Table 59

Combined Data on Size of Ice-Carrying Sector

Region	Number of structures considered	$B_{Л}/B_{п}$	Size l of span, m	Source
U.S.	9	0.44-0.72	9.2-19.5	[43]
Scandinavia	8	0.26-0.72	8.6-50.0	[43]
Germany & Switzerland	8	0.35-0.80	10.0-54.0	[43]
USSR				
Svir'	3	0.52	18.2-32.0	[43]
Dnepr	3	0.55	13.0-17.5	[43]
Volga	4		20.0	[180]
Yenisey	1		25.0	[179]
Ob'	1		20.0	[178]
Vilyuy	1		30.0	[148]
Angara	1		12.0 (crest)	[180]
Kama	1		12.0 (crest)	[180]
Irtysh	2		18.0	[207]

The placement in plane of the ice-carrying openings should be recommended within the limits of the stream centerline, while in case of location on a river bend, the openings should be placed at the concave shore.

We should make particular reference to the undesirability of setting up, within the limits of the ice-carrying front, of individual deep openings, which situation would lead to a distinct unevenness in the carrying capacity of the dam and to ice accumulations.

3. Depth of Overflowing Layer in Ice-Passing Openings

Examples of ice accumulations forming on a dam crest are listed above, together with an indication of the relative danger inherent in this occurrence. Such buildups can have a considerable depth (to 4-5 m), but a relatively small area, in connection with which they are soon eroded by the current and are dumped into the tailwater.

However, the design thickness of the dam crest (at design level of ice passage) should be greater than the maximal depth h of ice pack. In the opinion of N.N. Pavlovskiy [42], the minimal depth of the overflowing layer on a dam crest (see Fig.5) with allowance for the required reserve (20%) during dumping of the ice comprises

$$H = 0.9h + 0.2h = 1.1h \qquad (24.2)$$

or, with allowance for the form of decay curve:

$$H = 1.1h/0.65 = 1.7h. \qquad (24.3)$$

It should be noted that owing to the crowding, by the spillway crest, of the river's cross section, we have the passage of floes having piled onto each other in several layers (2-4 and more). In connection with this, it is important to have an adequate depth of flow in the ice-carrying openings, attaining up to 6-7 m under the conditions of a harsh climate.

4. Suggested Calculation Procedure

As we observe, efficient procedures have not yet been developed for establishing the width of the ice-carrying sector and the minimal dimensions of the spans. Until recently, extensive use has been made of fairly tenuous analogies with a river regime in a natural state and with the work experience obtained from a limited number of structures which have been built.

The broad undertaking of construction of hydraulic engineering facilities and bridges on the Siberian rivers, the ice regime of which has a number of unique features, requires a more profound study of the problem.

Let us consider a possible method for finding the minimally permissible size of a bridge span.

A river with width B_0 is traversed by a bridge, the opening of which is divided into (n+1) equal spans with a dimension (Fig. 69)

$$l = B_0/n+1. \qquad (24.4)$$

Our observations conducted on the Siberian rivers permit us to conclude that during a full-scale ice passage, the dimensions of the large floes occurring most often range from 1/20 to 1/8 of the river's width, where the floes have about the same width and length. We postulate further that the floes of such sizes move during the full-scale ice passage, contacting with each other in individual sectors, and seemingly represent a unified field of length L_0, with mass m, having the reserve of kinetic energy:

$$T = mv^2_0/2. \qquad (24.5)$$

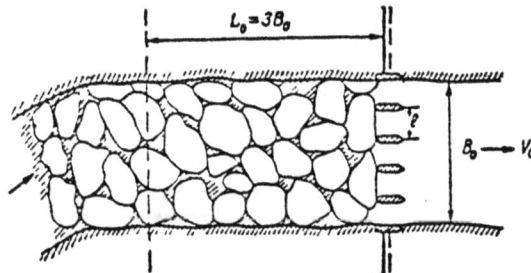

Fig. 69. On Determining the Value of Ice-Carrying Openings.

Studies of operation of ice-shielding floating barriers conducted by A.N. Latyshenkov [152] permit us to establish that at $L_0 > 3B_0$, further increase in pressure on the floating barrier does not occur owing to the fact that a part of the pressure is transmitted to the waterway shore.

Utilizing Eq. (24.5), it should be kept in mind that the velocity of floes during the full-scale ice passage can be assumed to equal about 0.9 of the surface current velocity. Let us note further that even during the full-scale ice passage, the floes cover only part of the river surface, in connection with which we can assume:

$$\Omega = pB_0 L_0 = 3pB^2_0,$$

where p = percentage of river surface covered with ice.

In this manner, Eq. (24.5) acquires the form

$$T = \frac{\varkappa v_0^2}{2} = \frac{h\Omega\gamma v_\pi^2}{2g} = \frac{B_0^2 h p v_0^2}{8,6},$$ (24.6)

and the floe's kinetic energy reserve being transmitted to one pier proves to equal

$$T_1 = \frac{B^2 h p v_0^2}{8.6n}$$ (24.7)

It is evident that jamming of ice does not occur at the structures if the kinetic energy reserve T_1 is sufficient for demolishing one floe with length L, closest to the pier.

Thus the inequation

$$T_1 > H_{max} L,$$ (24.8)

is valid, from which we can find

$$\lambda = \frac{B_0}{n} \geqslant \frac{8,6 H_{max} L}{h p v_0^2 B_0}.$$

It is easy to see that the value λ is always greater than span size l, wherein the equation

$$l = \frac{n}{n+1}\lambda.$$ (24.9)

is valid.

Assigning various values from 4 to 10 to the number n of piers, we can establish that the n/n+1-value will fluctuate from 0.8 to 0.91 and on an average can be assumed to equal 0.85. Then the minimally permissible size of span can be found from the expression:

$$l \geqslant \frac{7,3 H_{max} L}{h p v_0^2 B_0}$$ (24.10)

-288-

in which by H_{max} we signify the maximum horizontal pressure on a pier, developing at disruption of the ice pack from crushing, shearing or bending. As is known, this value can be found with one of the following expressions:

$$\text{during crushing......} \quad H_{max} = K_p b_o h,$$

$$\text{during shearing......} \quad H_{max} = K_c b_o h,$$

$$\text{during bending ......} \quad H_{max} = K_\text{и} b_o h$$

In this way, the minimally permissible value of a span can be found from the expression

$$l \geq \frac{7.3 K_{min} b_o L}{p v_0^2 B_0},$$ (24.11)

where K_{min} = the lowest of the K_p, K_c or $K_\text{и}$ -values.

Equation (24.11) takes into account current velocity v_o, ice passage density p, pier sizes b_o, dimensions of floes L/B_o and also the ice strength and form of pier in plane and profile, K_{min}, and thus reflects the influence of all the most significant factors.

Field observations of the ice passage process on the Siberian rivers have shown that the maximum density of ice passage can be estimated with the value $p = 0.7 - 0.85$. Taking into account further that some of the kinetic energy from the accumulation of floes under consideration is transmitted to the shores, we can recommend adopting the analogous conditions of ice-out:

$$p = 0.66.$$

The dimensions of the largest floes, occurring during the full-scale ice passage, for the conditions of Siberia comprise 1/8 - 1/20 of the river's width, i.e. from 30 to 100-150 m. In connection with this, we suggest that one can assume

$$L/B_o = 0.10.$$

With these raw data, Eq. (24.11) assumes the form:

$$l \geq \frac{1.1 K_{min} b_o}{v_0^2}.$$ (24.12)

Let us note that, considering the ice passage conditions during the full-scale ice-out, we can assume a somewhat reduced ultimate strength of ice to crushing, shearing and bending.

5. Examples of Calculation

1. Find the minimally permissible size of bridge span, proceeding from the conditions of unjammed ice passage.

Original data:

$$b_o = 3 \text{ m}; \quad R_{compres} = 40 \text{ t/m}^2; \quad \beta = 90°$$

$$v_o = 2/\text{sec}; \quad h = 0.75 \text{ m}; \quad \alpha = 45°$$

$$p = 0.66; \quad k = 0.50; \quad L/B_o = 0.10.$$

Since the pier has a vertical cutting face, disruption is possible only from crushing at

$$K_{min} = K_p = \frac{2.5 R_{compres} mk}{\sqrt[3]{v}} = 23.8 \text{ t/m}^2$$

and

$$l \geqslant \frac{7.3 K_{min} b_o L}{p v_o^2 B_o} \geqslant 19.8 \text{ m}.$$

2. Find the minimally permissible size of bridge span under conditions of the preceding example but in the presence of a sloped ice-cutting face and the following additional data:

$$\beta = 60°; \quad R_{sh} = 20 \text{ t/m}^2; \quad R_и = 40 \text{ t/m}^2.$$

In this case, demolition can occur from shearing or bending. In disruption by shearing, we have:

$$K_{sh} = 1.1 R_{sh} k \frac{\tan \beta}{\sin \alpha} = 27.0 \text{ t/m}^2.$$

In the case of breakdown from bending, we similarly find (11.43)

$$K_и = R_и \tan \beta \, S_o = 40 \times 1.73 \times 0.18 = 12.5 \text{ t/m}^2$$

In the given case, the disruption will thus occur from bending and the minimally permissible size of span can be assumed to equal

$$l \geqslant \frac{7.3 \cdot 12.5 \cdot 3.0 \cdot 10}{0.66 \cdot 2} \geqslant 10.1 \text{ m}$$

As we shall see, the application of a sloped ice apron facilitates the destruction and permits a smaller size of spans.

Table 60 and Fig. 70 permit us to estimate the value of the minimally permissible span at various values of K_{min} and v_0.

Table 60

Values of Minimal Span Size, Meters

K_{min}, t/m^2	v_0, m/sec				Remarks
	1.5	2.0	2.5	3.0	
15	22.5	12.4	7.9	5.6	At b_0 = 3 m
30	45.0	24.7	15.7	10.3	
45	67.5	36.3	23.6	16.8	

The table is compiled for b_0 = 3 m but can also be utilized for piers of another width, for which it is necessary to multiply the derived span value by ratio $b_0/3$.

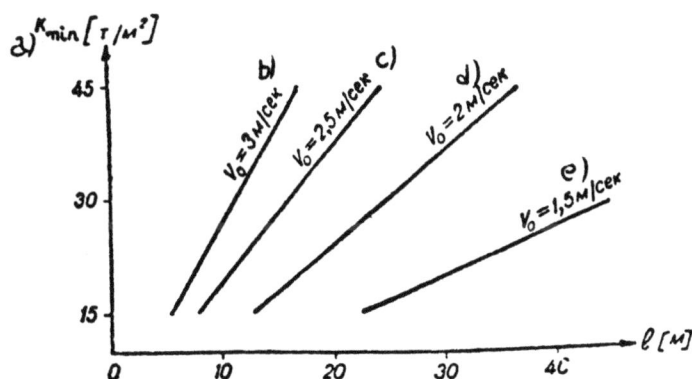

Fig. 70. Graph for Finding Value of Ice-Carrying Openings. Key: a) K_{min} [t/m^2]; b) v_0 = 3 m/sec; c) v_0 = 2.5 m/sec; d) v_0 = 2 m/sec and e) v_0 1.5 m/sec.

6. Brief Conclusions

1. In the division of a bridge opening into spans, it is necessary to take into account (in addition to the concepts of a technical-economic nature and the requirements imposed by shipping or timber rafting) the necessity for the unjammed passage of ice.

2. The stopping of floes at the structure in the initial period of ice passage is quite permissible, since the subsequent rise in water levels and the approach of new large ice fields facilitates the breakup of the floe which had stopped previously.

3. The value of the individual spans should not be assumed less than the value:

$$l \geqslant \frac{7.3 \ K_{min} b_o L}{p v_o^2 B_o} \qquad (24.11)$$

4. With frequently occurring values of L/B_o and p, Eq. (24.11) can be given a simpler form:

$$l \geqslant \frac{1.1 K_{min} b_o}{v_o^2}$$

5. In dividing the ice-carrying sector of spillway dams, the size of a span can be assumed somewhat smaller (by 10-20%) owing to the effect of the floes' breakup on the decay curve.

EIGHTH PART

FINDINGS AND CONCLUSIONS

Chapter XXV

BASIC CONCLUSIONS AND RECOMMENDATIONS

At the end of almost every chapter in this book, we have
given detailed conclusions, summarizing the results of reviewing
each given question.

We will discuss below only the essential conclusions,
generalizing the results of the field observations conducted and
also of the theoretical and experimental research.

They can be formulated briefly in the following manner.

1. The extensive launching of the construction of new
bridges, hydraulic engineering and water-storing facilities
(many of which are subjected to the action of the ice pack)
makes very urgent the solution to the problem of combatting the
ice both during the winter and particularly during the spring
ice passage. A consideration of the action of the ice pack on
structures is necessary since for many USSR rivers, we typically
have a great thickness and strength of the ice.

2. The solid structures of great extent (dams, embankments, arches), having a significant mass, are capable of withstanding the ice action successfully. The separately standing
structures (bridge piers, individual supports of spillway dams,
water-storing facilities, beacons) are in more difficult circumstances.

3. Across the vast territory of the Soviet Union, there
are regions differing quite markedly in conditions of the spring
ice-out of flowing rivers. We can differentiate:

a) rivers with severe conditions of spring ice-out, typified by a great thickness (more than 0.8 m) and strength of ice,
by the presence of large ice fields, by a significant rise in
water level and the frequent formation of ice jams (middle and
lower reaches of the large Siberian rivers and also certain
rivers in the European North);

b) rivers with an intensive spring ice passage, typified by an appreciable ice thickness (0.4 - 0.8 m), by the presence of large floes during the ice passage, sometimes occurring with the formation of jams (upper reaches of the medium sized and large rivers in Siberia, middle course of the southward-flowing rivers in the European part of the USSR, certain rivers in the Far East); and

c) rivers with slight spring ice passage, characterized by slight thickness (less than 0.4 m) and strength of ice, by a slight rise in water level, and by small dimensions of floes during the full-scale ice passage (rivers in the South of the European part of the USSR, rivers in Central Asia).

4.   Dynamic ice pressure is most significant during the spring ice passage, especially for the individually standing structural supports. The static pressure developing during thermal expansion has significance chiefly for the design of dams, embankments, arches and other structures of great extent.

5. Observations of the operation of structures during the ice passage period permitted us to establish that the process of demolishing an ice pack begins from local deformations of the floes in the contact area. The developing force of interaction gradually increases and finally attains a value sufficient for: a) stopping the floes at the pier or the structural sector; b) splitting the floe; c) cutting through the floe without splitting it; or d) demolishing the floe from shearing or bending.

The nature of the breakdown is established mainly by the reserve of the floe's kinetic energy and also by the pier's form and dimensions.

6. Close contact of the ice pack with the pier is possible only with a weak ice sheet, not dangerous per se. With a strong ice pack, the width of the pier part absorbing the ice pressure does not exceed 0.4 - 0.8 of the maximum pier width. In connection with this, the suggestion has been made to introduce into the calculations a completeness factor of contact of pier and floe, less than unity.

7. The elastic and plastic properties of ice have still been little studied owing to the diversity in the behavior features of an ice pack under load (relationship of deformations with temperature, time of load effect, its value, ice structure, recrystallization processes, etc.).

8. Ultimate ice strength during compression, crushing, bending, shearing and tension is connected with the ice temperature, its structure, orientation of force relative to the crystallization axis, rate of deformation and with other test conditions.

We can conclude that, up to the time of the spring ice passage, the ultimate compressive and bending strength of river ice decreases considerably; a very significant effect on the ice's resistance to the penetration of a pier is exerted by the local crushing phenomenon, capable of increasing the developing interaction forces by 2 - 2.5 times; an appreciable role is played by the deformation rate; with its increase, the ultimate compressive and crushing strength of ice decreases, being subjected to the relationship:

$$R_{compres} = R/\sqrt[3]{S}.$$

A similar pattern has also been detected in determining the bending strength of ice. Of great importance are the dimensions of the sample; with their increase, the ultimate ice strength decreases. At the suggestion of I.P. Butyagin, one should assume the relative bending strength for blocks (8 X 8 cm) to be 1:00; for slabs (50 X 80 cm), to be 0.33-0.50, and for ice fields, 0.25-0.35.

9. The values for the ultimate strength of ice (for temperatures close to 0°C and a definite orientation of force relative to the crystallization axis) are given in Table 61.

Table 61

Calculated Values of Ultimate Ice Strength, t/m^2

Nature of stresses	Direction of force	Symbol	Rivers of North&Siberia			Rivers of European part of USSR		
			thrust, $v=0.5$ m/sec	full-scale ice-out		thrust, $v=0.5$ m/sec	full-scale ice-out	
				$v=1$ m/sec	$v=1.5$ m/sec		$v=1$ m/sec	$v=1.5$ m/sec
Compression	perpendic	$R_{c \perp}$	65	50	45	40	30	25
Local crushing	"	R_{cm}	150	125	110	80	65	55
Bending		$R_{\shortparallel}$	65	50	45	40	30	25
Shearing	parallel	R_{sh}	--	40-60	--	--	20-30	--
Tension	"	R_p	--	70-90	--	--	30-40	--

10. The experiments undertaken in finding the ice resistance to the penetration of punches demonstrated the great effect of the punch's shape (and hence of a pier's form in plane)on the resistance of an ice pack. Thus, utilizing in place of a semicircular configuration a triangular one with a tapering angle of 60°, we observed a reduction of 1.5 - 2 times in the force required for the introduction of the punch. The effect of the pier's shape can be taken into account by introducing a form coefficient m into the calculations for which we recommend numerical values.

11. The ice pressure developing during cutting by a solid pier (with a vertical cutting face), of a large ice field is most likely and is computed with the equation:

$$P = K_p b_o h,$$

where b_o = width of pier, h = ice thickness; and K_p = coefficient taking local conditions into account and equalling:

$$K_p = \frac{2.5 \, R_{c*} mk}{\sqrt[3]{v}}$$

[Trans. Note: Here et passim, subscript "c_*" = "compressive"]

This equation allows for the pier's form m in plane, nondense state of osculation k, floe velocity v, ice strength R_{c*}, local crushing pehnomenon (2.5) and refines to a considerable extent the recommendations previously suggested.

12. The ice pressure developing during the stopping of an ice field at a pier can be found with the formula

$$P = 0.68vh \sqrt{\Omega R_{c*} \, m \tan \alpha},$$

taking into account the pier's form in plane, m, α, floe dimensions Ω, h, their rate of movement v and ice strength $R_{c\,M}$.

The equation is given for piers with a triangular. shape of cutting face in plane but can also be extended to the piers with a semicircular outline (which are equivalent to the triangular one with tapering angle $2\alpha = 140°$).

13. The forces developing during the splitting of small floes can be determined with the following equations:

at splitting on the side

$$P = 2R_{sp}Bh \tan \measuredangle /2,$$

at partial splitting

$$P = 2R_{sp}hS \sin \measuredangle ,$$

during cleavage

$$P = nR_pLh,$$

where L, B, and h = length width and thickness of floe; n = coefficient taking into account the pier's form in plane, the ice strength and current velocity.

14. An analysis of the equations derived has shown that during the time of ice thrusts and full-scale ice passage, the appearance of a maximal pressure equalling

$$P = K_p b_o h$$

is almost always possible, which also should most often be the design value.

15. In case of the sloping of the ice-cutting face to the horizon, the latter as it were cuts the ice field from beneath, which promotes a more successful breakdown of the ice pack and reduces the possibility of the formation of ice jams. Demolition of a floe upon contact with the sloped face of an ice apron can occur from bending or shearing for strong or weakened ice, respectively.

16. Considering the significant increase in the scale and cost of operations caused by the installation of an inclined face of solid piers, one should resort to such a step only in extreme cases, preceded each time by a careful justification. Proceeding from the experience gained in the operation of piers, we can conclude that under the USSR conditions, one can almost always get by without the use of a sloped ice-cutting face. In all cases, the wooden starlings should be provided with a sloped cutting face, permitting them to break the ice pack more successfully.

17. During the ice action on a sloped ice apron, the horizontal H and vertical V components develop, interrelated at any moment by the relationship:

$$H = V \tan \beta_1,$$

where β_1 = inclination angle of the lateral edge faces to the horizon.

In the case of disintegration from shearing or bending, value V_{max} of vertical pressure component is determinative. In this case, the horizontal component can be found with the formula:

$$H_{max} = 1.1 V_{max} \tan \beta ,$$

where coefficient 1.1 takes into account (according to B.V. Zylev's suggestion) the friction between the pier and the floe.

18. The analysis conducted on the conditions of the break down of an ice pack from shearing and bending permitted the suggestion of the following procedures for finding the ice pressure on piers in this instance.

At disruption from shearing

$$H_{cp} = 1.1 b_o h R_{cp} k \tan \beta / \sin \alpha = K_{cp} b_o h ,$$

at disruption from bending

$$H_и = R_и h^2 \tan \beta \; b_o / h \; S_o = K_и b_o h,$$

where the lowest of them should comprise the design value.

The suggested formulas take into account ice strength R_{sh}, $R_и$ (subscript "и" = "bending"), its thickness h, nondense state of osculation k, plus dimensions b_o and the form of a pier in plane and profile α, β, and S_o.

19. The working formulas for crushing, shearing and bending are compared with 18 field observations, which permits their recommendation for the practice.

20. The form of a pier in plane and profile has a very significant effect on the interaction forces developing between a pier and a floe.

The inclination of the cutting face of an ice apron leads to a considerable decrease in the horizontal component of the ice pressure on a pier. Excessively steep ice aprons (at $\beta > 75$-$80°$) do not yield the expected effect and therefore should not be utilized.

In the piers with a vertical cutting face, sharper outlines of the pier's form in plane are extremely desirable. Thus,
modifying the pier's tapering angle from 120° to 60°, we obtain
a reduction in the force of ice pressure:

      at cutting through a large floe... by 1.33 times

      at stopping of a floe .... by 2.00 times

      at splitting of a floe .... by 2.10 times

We should thus avoid the blunt outlines of a cutting face
in plane and make more extensive use of a pier with minimally
possible (based on design concepts) tapering angles.

21. Consideration of the piers' deformations is logical
only in the case of the stopping of small floes at piers of
slight rigidity with pliability factor a > 0.001. In the case of
the disruption of large floes and also in the application of
solid piers, the allowance for the deformations does not have
practical meaning.

22. For the correction of the proposed formulas, the under
taking of a determination of ice pressure under field conditions
is quite desirable. However, the methods having existed until
now for finding the ice pressure in nature have required expensive and complex equipment and hence can not find broad application in practice.

23. An analysis of the movement of a large floe being cut
by a pier provided the opportunity of obtaining formulas linking
the floe velocity V_η , current velocity V_o, floe's dimensions
Ω_l, h, M, roughness f of their lower surface, length s of cut-
through sector of ice field with time t, spent for the cutting.

As a result of the utilization of these formulas, a kinematic method has been offered for finding ice pressure in nature,
developing during the cutting of a large floe.

24. This means of finding the ice pressure value under
field conditions is quite simple and permits one, without complex equipment, to establish it for piers of various dimensions,
form and material.

It is advantageous to use filming of the process of a
large floe's breakdown by a pier, which also makes it possible to
obtain all the necessary data for calculation and for the
special equipment which has been developed.

25. The kinematic method has been used in 18 cases for finding ice pressure under field conditions in the rivers of Siberia. A comparison of the results with the calculations based on the theoretical formulas and standards has indicated their satisfactory agreement and has afforded the opportunity of critically evaluating the existing Technical Specifications for establishing the ice loads (SN 76-59).

26. In the book, conditions have been reviewed of ice passage through a structural site and a technique has been proposed for finding the minimally permissible size of a bridge span, proceeding from the requirement for the unjammed passage of ice.

27. In the book, a critical examination has also been made of the methods of determining the ice action in other cases, specifically:

a) during fluctuations in water level from the frozen ice pack;

b) during drift of an ice field caused by current or wind;

c) during dumping of ice into the tailwater;

d) during ice accumulations in a structure;

e) during the abrasive action of ice; and

f) during dynamic pressure of ice on extensive structures.

Certain recommendations have been expressed in counter-acting the harmful effect of ice.

28. Thus, as a result of the present report and a gener-alization of the studies undertaken by other authors, the basic questions of ice action on the piers of bridges and of hydraulic engineering facilities have gained, as it were, a known founda-tion for their final solution.

The pattern has been clarified of the functioning of structures during the spring ice passage; a basic study has been made of the mechanical properties of ice and the effect of various factors on them; methods have been developed for finding the ice pressure with vertical and sloped cutting faces.

In addition, we have established the effect of a pier's form in plane in profile, and we have also suggested simple techniques for finding ice pressure under field conditions. A formula has been offered, permitting a determination of the minimally permissible size of the ice-carrying openings of dams and bridge spans.

It is nevertheless necessary to point out that certain questions of each action on structures are still unsolved and thus the undertaking of further research is necessary.

Among the priority questions which in our view should be solved during the additional research, one should include:

a) organization of further observations in nature of the action of large floes on structures, with the aid of the kinematic method;

b) the undertaking of field observations of the value of static ice pressure on structures of great extent;

c) the generalization and thorough analysis of the operation of the completed structures, as well as those being built, on rivers with considerable ice passage;

d) an additional study of the physico-mechanical properties of an ice pack and especially of the inherent tendencies linking the ice deformation, effective stresses, acting time of load and ice temperature.

e) development of a procedure for simulating the ice processes.

A thorough study of the processes of ice action on structures in nature will permit the selection of the more substantiated design systems, the evaluation of the possibility of introducing some given assumptions and a conduct of further refinements in the methods used for assessing the ice action on engineering structures.

We also need a closer coordination in the work of groups functioning in the field of ice thermics, coupled with the organization of a broad exchange of the research findings.

BIBLIOGRAPHY

[1]   Komarovskiy, A.N., Deystviye ledyanogo pokrova na sooruz-
      heniya i bor'ba s nim (Effect of an Ice Pack on Structures
      and Combatting It), Part 1, Gosenergoizdat, Moscow, 1932.

[2]   Bliznyak, Ye. V., "Materials on Description of Russian
      Rivers," Part 11, No 9, St. Petersburg, 1916.

[3]   Bliznyak, Ye. V., and Polyakov, B.V. Inzherernaya gidrologiya
      (Engineering Hydrology), Moscow-Leningrad, 1939.

[4]   Timonov, V.Ye., "Ice Engineering as Combination of Scientific
      Disciplines of Great Significance for the USSR National
      Economy," Izv. VNIIG, No 2.

[5]   Bovin, V.T., "Ice and Its Counteraction in Hydraulic Engineer
      ing Structures," Moscow, 1927.

[6]   Bydin, F.I., "Ice Difficulties and Struggle with Them. Izv.
      VNIIG, No 8, Leningrad, 1932.

[7]   Bydin, F.I., "Ice Passage and Complications Associated with
      It," Priroda (Nature), No. 2 Moscow, 1949.

[8]   Artamonov, D.S., "Collection on Ice Engineering," No 1
      Gosstroyizdat, Moscow-Leningrad, 1933.

[9]   Antonov, N.D., Trudy Arkticheskogo Instituta (Transactions
      of Arctic Institute), Vol 105, Leningrad, 1938.

[10] Vasenko, B.P., "Ice," Sb. In-ta inzh. put. soobshch., No 1,
      St. Petersburg, 1899.

[11] Makarov, S.O., Yermak vo l'dakh (The Yermakh in the Ice),
      St. Petersburg, 1901.

[12] Veynberg, B.P., Led (Ice), Gostekhteoizdat, Moscow-Leningrad,
      1940.

[13] Veynberg, B.P., "Results of Measuring Specific Weight and
      Strength of Ice on Tomi R. Prior to 1912 Ice-out. Izv.
      Tomsk. Tekh. In-ta (Bull. of Tomsk Engineering Institute),
      Tomsk, 1913.

[14] Pinegin, V.N., "Preliminary Report on Investigating the Strength of River Ice in Connection with Temperature Variations," Vestnik Sib. Inzh. (Herals of Siberian Engineers), Vol 4, Tomsk, 1924.

[15] Sergeyev, B.N., "Construction of Winter Crossing Over Ice and Function of Ice Layer Under a Load," Sb. Inzh. Issledovaniy NKPS, No 18.

[16] Basin, M.M., Ispytaniye ledyanogo pokrova r, Sviri na prochnost' pri szhatil, skalyvanii i izgibe v period tananiya vesnoyu 1934 goda (Testing of Ice Sheet on Svir' R. for Compressive, Shearing and Bending Strength in Thawing Period, Spring 1934), Izd. NIIG, Leningrad, 1935.

[17] Nazarov, V.S., "Property of Ice and Its Trafficability by Ships," Morskoy Sbornik (Marine Collection), No. 11-12,1941.

[18] Vitman, F.F., and Shendrikov, N.P., "Certain Investigations of Mechanical Properties and Strength of Ice," Trudy Arkticheskogo In-ta, Vol 110, 1938.

[19] Korzhavin, K.N., "Investigation of Mechanical Properties of River Ice," Tr. NIIZHT, Vol. 4, Novosibirsk, 1940.

[20] Korzhavin, K.N., "Observations of Variation in Ice Strength on Ob' R. by Period of Spring Ice Debacle in 1934." Tr. NIIZHT, No 111, Tomsk, 1938.

[21]. Korzhavin, K. N., "Effect of Deformation Rate on Ultimate Compressive Strength of River Ice," Tr. NIIZHT, No. 11, Novosibirsk, 1955.

[22] Korzhavin, K. N., "Effect of Local Crushing on Mechanical Properties of River Ice," Tr. TEI SO AN SSR (Transactions of Ice Thermics Institute of USSR Academy of Sciences, Siberian Branch), No 11, Novosibirsk, 1961.

[23] Izyumov, S.M., "Ice Crossings," Tekhnika i Vooruzheniye (Engineering and Armament), No 1-2, 1942.

[24] Kartashkin, B.D., "Experimental Study of Physico-Mechanical Properties of Ice," Trudy TSAGI, No 607, Moscow, 1947.

[25] Shul'man, A.R., Khanina, S.K., Fillippova, L.I, and Donchenko, R.V., Trudy GGI (Transactions of State Hydrologic Institute), No 6 (70), Leningrad, 1949.

[26] Berdennikov, V.P., "Experiment in Calculating Elasticity of Spring Ice," Tr. GGI, No 6 (70), Leningrad, 1949.

[27] Zubov, N.N., Morskiye Vody i l'dy (Sea Water and Ice), Izd. GUSMP, Leningrad, 1938.

[28] Neronov, Yu. N., "On the Question of Amount of Temporary Bending Resistance of Spring Thawing Ice," Tr. NIU GUGMS, Ser. 5, No. 20, Gidrometeoizdat (Hydrometeorological Press), Moscow, 1946.

[29] Tsytovich, N.A., Osnovy mekhaniki gruntov (Fundamentals of Soil Mechanics), OGIZ, Moscow, Leningrad, 1940.

[30] Butyagin, I.P., "Strength of Ice Cover in Ice Loads on Hydraulic Structures," Tr. TEI SO AN SSSR, No. 11, Novosibirsk, 1961.

[31] Korzhavin, K.N., and Butyagin, I.P., "Investigation on Deformations and Strength of Ice Fields Under Natural Conditions," Tr. TEI SO AN SSSR, No 11, Novosibirsk, 1961.

[32] Butyagin, I.P., "On Strength of Ice Pack During Shearing Forces," Tr. TEI SO AN SSSR, No. 7, Novosibirsk, 1958.

[33] Voytkovskiy, K.F., Mekhanicheskiye svoystva l'da (Mechanical Properties of Ice), Moscow, 1960.

[34] Vyalov, S.S., "Dependence Between Stresses and Deformation of Frozen Soils with Allowance for Time Factor," Doklady Akademii Nauk SSSR (Proceedings of USSR Academy of Sciences), Vol 108, No 6, 1956.

[35] Vyalov, S.S., Reologicheskiye svoystva i nesushchaya sposobnost' merzlykh gruntov (Rheological Properties and Carrying Capacity of Frozen Soils), Moscow, 1959.

[36] Lavrov, V.V., "Nature of Scale Effect in Ice and Strength of Ice Sheet," Doklady Akademii Nauk SSSR, Vol 122, No 4, Moscow, 1958.

[37] Savel'yev, B.A., Izucheniye mekhanicheskikh i fizicheskikh svoystvl'da (Study of Mechanical and Physical Properties of Ice), Moscow, 1957.

[38] Peschanskiy, I.S., "On Compression of Vessels and Hummocking Forces," Problemy Arktiki (Problems of the Arctic), No 3, 1944.

[39] Perederiy, G.P., _Kurs mostov_ (Course on Bridges), Moscow, 1945.

[40] Nikolai, L.F., _Ob opredelenii poperechnykh razmerov bykov v zavisimosti ot ledokhoda_ (On Determining the Transverse Dimensions of Piers Depending on Ice Passage), _Izd. Sobr. Inzh. Putey Soobshch._, 1897.

[41] Shchapov, N.M., "Impact of Floes on Structures," _Gidro-tekhnicheskoye Stroitel'stvo_ (Hydraulic Engineering Construction), No. 2 1933.

[42] Rynin, N.A., _Ledorezy_ (Ice Aprons), _Izd. Inta. Inah. Putey Soobschch._, St. Petersburg, 1903.

[43] Komarovskiy, A.N., _Deystviye ledyanogo pokrova na sooruzheniya i bor'ba s nim_ (Action of an Ice Cover on Structures and Ways to Combat It), Part 2, Moscow, 1933.

[44] Bydin, F.I., "Development of Certain Questions in Area of Rivers' Winter Regime," _Tr. III. Gidrol. S'yezda_ (Trans. of Third Hydrologic Congress), Leningrad, 1959.

[45] Zylev, B.V., "Ice Pressure on Sloped Ice Aprons," _Tr. MIIT_, No. 74, _Transzheldorizdat_, Moscow, 1950.

[46] Gamayunov, A.I., "Determination of Ice Pressure on Bridge Piers Under Natural Conditions," _Tekhnika Zheleznykh Dorog_ (Railway Engineering), No 4 and 12, 1947.

[47] Kuznetsov, P.A., _Deystviye l'da na sooruzheniya morskikikh portov i zashchita ot nego_ (Effect of Ice on Structures of Seaports and Defense Against It), Leningrad, 1939.

[48] Petrunichev, N.N., _O dinamicheskom davlenii l'da_ (On the Dynamic Pressure of Ice), In the book: "Ice Thermal Problems in Hydraulic Power Engineering," Leningrad, 1954.

[49] Dubakh, A.A., "Life of a River," _Izv. Goretskogo SKHI_, 1925.

[50] Platonov, Ye.V., _Opory mostov_ (Bridge Piers), Moscow, 1946.

[51] Kuznetsov, P.A., "Ice Loads on Hydraulic Engineering Structures." _Sbornik trudov LONITOVT_, Leningrad, 1948.

[52] Konovalov, I.M., Orlov, P.N., and Yemel'yanov, K.S., Osnovy ledotekhniki rechnogo transporta (Fundamentals of Ice Engineering in River Transport), Moscow-Leningrad, 1952.

[53] Korzhavin, K.N., Davleniye l'da na opory mostov i gidro-tekhnicheskikh sooruzheniy (Ice Pressure on Supports of Bridges and Hydraulic Engineering Structures, Novosibirsk, 1939.

[54] Korzhavin, K.N., "Operation of Ice Aprons Around Bridge Piers Under Ice Passage Conditions in Siberian Rivers," Tr. NIVIT, No 3, 1938.

[55] Korzhavin, N.N., "Action of Ice on Supports of Bridges and Hydraulic Engineering Structures," Tr. TEI ZSFAN SSSR, No 5, Novosibirsk, 1955.

[56] Korzhavin, N.N., "On an Efficient Type of Bridge Support Under Conditions of Ice Passage in Siberian Rivers." Tr. NIVIT, No 7, 1949

[57] Korzhavin, K.N., Vozdeystviye l'da no opory mostov (Action of Ice on Bridge Piers), Moscow, 1951.

[58] Korzhavin, K.N., "Effect of Slope Angle of Ice-Cutting Face of Bridge Support on Amount of Ice Pressure," Tr. TEI ZSFAN SSSR, No 5, Novosibirsk, 1955.

[59] Skryl'nikov, V.P., Plotiny gravitatsionnyye (Gravity Dams), Moscow, 1933.

[60] Anisimov, N.I., Vodokhranilishchnyye plotiny (Reservoir Dams), OGIZ, Moscow, 1931.

[61] Golushkevich, S.S., O nekotorykh zadachakh teorii izgiba ledyanogo pokrova (On Certain Problems in the Theory of Bending the Ice Sheet), Leningrad, 1947.

[62] Petrunichev, N.N., O staticheskom davlenii l'da (On the Static Pressure of Ice) In the book: "Ice Thermal Problems in Hydraulic Engineering" Leningrad, 1954.

[63] Royen, N., Istrick vid temperatur högningar [in Swedish], Stockholm, 1922.

[64] Proskuryakov, B.V., "Static Pressure of Ice on Structures," <u>Tr. GGI</u>, No 4 (58), 1947.

[65] Girillovich, N.A., "Planning the Ice Regime of Low-Pressure River Channel Hydraulic Stations," <u>Izv. NIIG</u>, No 22,1938.

[66] Bibikov, D.N., and Petrunichev, N.N., <u>Ledovyye zatrudneniya na gidrostantsiyakh</u> (Ice Difficulties at Hydraulic Stations) Moscow-Leningrad, 1950.

[67] Petrov, G.N., and Trufanov, A.A., "Eliminate the Contradictions in Instructions for Designing Ice Aprons." <u>Stroitel'stvo dorog</u> (Road Building), No 8-9, 1940.

[68] Karlsen, G., and Streletskiy, N., "Wooden Ice Aprons," <u>Voyenno-inzherernyy Zhurnal</u> (Military Engineering Journal), No 1, 1945.

[69] O'mezov, A.S., and Shpiro, G.S., "Deformation of High Pile Gratings of One Bridge," <u>Tr. VNIIZHT</u>, No 26, Moscow, 1948.

[70] Gibshman, Ye.Ye., <u>Derevyannyye mosty na avtomobil'nykh dorogakh</u> (Wooden Bridges Along Highways), <u>Izd. Narkomkhoza RSFSR</u>, Moscow, 1938.

[71] Boldakov, Ye.V., <u>Perekhody cherez bol'shiye vodotoki</u> (Crossings Over Large Waterways), <u>Dorizdat</u>, Moscow, 1949.

[72] Sukhorukov, A.Ya., "Design Features of Icebreakers' Hull, <u>Tr. Gidrogr. in-ta GUSMP</u>, No 2, Leningrad, 1942.

[73] Maslov, A.I., "Experience in Calculating the External Forces Acting on a Ship Hull under Ice Conditions," <u>Tr. VNITOSS</u>, Vol 3, No 3, Moscow-Leningrad, 1937.

[74] Bovin, V.T., "Observations of Spring Flood in 1926 at Volkhovstroy, " Project Materials of Prof. I.G. Aleksandrov, No 3, Moscow, 1927.

[75] Katkova, S.A., "On the Advancement of Ice on the Upper Slope of an Earth Dam," <u>Gidrotekhnicheskoye Stroitel'stvo</u> (Hydraulic Engineering Construction), No 7, 1961.

[76] Samochkin, V.N., "Action of Ice Passages on Shore Slopes and Structures," <u>Tr. TEI SO AN SSSR</u>, No 11, 1961.

[77] Sokol'nikov, V.N., "Features in Formation, Accretion and Breakdown of Ice Pack on Lake Baykal," <u>Tr. III Gidrol. S'yezda</u>, Leningrad, 1959.

[78] Kuzub, G.Ya., "Temperature Cracks in Ice Pack," <u>Tr. TEI SO AN SSSR</u>, No 7, 1959.

[79] Kirkham, I.E., "Design of Ice Breaker Nose for Missouri River Bridges," Eng. News Record, No 9, 1929.

[80] Perederiy, G.P., Kurs zhelezobetonnykh mostov (Course on Ferroconcrete Bridges), <u>OGIZ</u>, 5th Ed., Moscow, 1931.

[81] Gel'fer, A.A., <u>Razrusheniye mostovykh opor i mery ikh zashchity</u> (Destruction of Bridge Piers and Measures to Protect Them), Moscow, 1938.

[82] Shafir, I.N., and Ginsberg, R.I., <u>Avarii morskikh gidrotekhnicheskikh sooruzheniy</u> (Damages to Marine Hydraulic Engineering Structures), Moscow, 1942.

[83] Zubov, N.N., <u>L'dy Arktiki</u> (Arctic Ice), Moscow, 1945.

[84] Timonov, V. Ye., and Tsionglinskiy, M.F., "Building of Dams on Rivers with Wide Variations in Output of Water and Intensive Ice Passages," <u>Tr. XI Mezhdunarod. sudokhodn. kongressa</u> (Transactions of 11th International Congress on Navigation), St. Petersburg, 1908.

[85] Gel'fer, A.A., <u>Prichiny i formy razrusheniya gidrotekhnicheskikh sooruzheniy</u> (Causes and Forms of Damages to Hydraulic Engineering Structures), Moscow, 1936.

[86] Ryabukho, A.M., "On the Question of Planning the Piers to Bridges Across Reservoirs," <u>Tr. NIIZHT</u>, no 27, Novosibirsk, 1961.

[87] Butyagin, I.P., "Thickness and Streucture of Ice Pack on Ob' R." <u>Tr. TEI SO AN SSSR</u>, No 7, 1958.

[88] Bydin, F.I., "Study of Ice Accretion under Natural Conditions," <u>Izv. VNIIG</u>, Vol 4, Leningrad, 1932.

[89] Piotrovich, V.V., "On Ice Thickness at Betinning of Ice Jamming in the Plains Rivers of ETC of USSR," <u>Tr. TSIP</u> (Transactions of Central Forecasting Institute), No 5, (32), 1947.

[90] Appolov, B.A., Ucheniye o rekakh (Study of Rivers), Izd. Mosk. un-ta (Moscow University Press), 1951.

[91] Syrnikov, P.I., Sposob izmereniya tolshchiny l'da bes probivaniya lunok (Methods of Measuring Ice Thickness Without Drilling Holes), Collection: "For Efficiency in Hydrology", 1933.

[92] Zaykov, B.D., Ocherki po ozerovedeniyu (Outlines of Limnology), Leningrad, 1955.

[93] Kritskiy, S.N., Menkel', M.F., and Rossinskiy, K.I., Zimniy termicheskiy rezhim vodokhranilishch, rek i kanalov (Winter Thermal Regime of Reservoirs, Rivers and Canals), Gosenergoizdat, Moscow-Leningrad, 1947.

[94] Voytkovskiy, K.F., Mekhanicheskiye svoystva l'da (Mechanical Properties of Ice, Moscow, 1960.

[95] Shumskiy, P.A. Osnovy strukturnogo ledovedeniya (Fundamentals of Structural Cryology), Moscow, 1955.

[96] Shumskiy, P.A., "The Mechanism of Ice Straining and Its Recrystallization," Symposium of Chamonix, Belgium, 1958.

[97] Lin'kov, Ye. M., "Elastic Properties of Ice and Methods for Studying Them." Vestnik Leningradskogo Universiteta (Herald of Leningrad University), No 6, Issue 3, Leningrad, 1957.

[98] Erwing, M., Crary, A.P., and Thorne, A.M., "Progatation of Elastic Waves in Ice., Physics, Vol 5, No 6, 1934.

[99] Pinegin, V.N., "On Variations in Elastic Modulus and Poisson Coefficient in River Ice During Compression," Nauka i Tekhnika (Science and Technology), No 3-4, 1927.

[100] Klyucharev, V., and Izyumov, S., "Determination of Load Capacity of Ice Crossings, VIZH, No 2-3, 1943.

[101] Lavrov, V.V., Doklady Akademii Nauk SSSR (Proceedings of USSR Academy of Sciences), Vol 122, No 4, 1958.

[102] Lavrov, V.V., "Ice Viscosity Depending on Temperature," Tekhnicheskaya Fizika (Technical Physics), Vol 17, No 9, 1947.

[103] Veynberg, B.P., "On Internal Friction of Ice," Zhurnal Russkogo Fiziko-Khimicheskogo Obshchestva (Journal of Russian Physico-Chemical Society), Vol 38, No 3-6, 1906.

[104] Petrunichev, N.N., Staticheskoye davleniye ledyanogo pokrova no gidrotekhnicheskiye sooruzheniya (Static Pressure of Ice Pack of Hydraulic Structures),Leningrad,1957.

[105] Bessonov, Ye., "Investigation of Ice on Tomi R. Prior to 1913 Ice Passage," Izv. Tomsk. tekhn. in-ta (Bulletin of Tomsk Engineering Institute), Tomsk, 1913.

[106] Pedder, Yu. A., "Observations of Strength of Ice on Angara River," Zhurnal geofiziki i meteorologii (Journal of Geophysics and Meteorology), Vol 6, Moscow, 1929.

[107] Arnol'd-Alyab'yev, V.I., "Studies of Ice Strength in the Gulf of Finland in 1923, 1927 and 1928," Izvestiya GGO, No 2, 1929.

[108] Yegorov, K. Ye., "Distribution of Stresses and Displacements in Dual-Layer Base of Banded Foundation," Sb. tr. nauch.-issl. sektora tresta glubinnykh rabot, No 10, Moscow-Leningrad, 1939.

[109] Gorbunov, Posadov, M.I., Osadki fundamentov na sloye grunta, podstilayemom skal'nym osnovaniyem (Settlings of Foundations of a Soil Layer Supported by a Rock Base), Stroyizdat, Moscow, 1946.

[110] Korotkin, V.G., "Volumetric Problems for Elastic-Isotropic Space," Sb. GIDEP, No 4, Leningrad, 1938.

[111] Sokolovskiy, V.V., Teoriya plastichnosti (Theory of Plasticity), 2nd Ed. Moscow-Leningrad, 1950.

[112] Nekrasov, S.F., "Operation of an Ice Road Under Load," Zapiski GGI (Notes of SHI), Vol 15, 1936.

[113] Pisarev, P.A., "Operation of Ice When Bending," Dissertation, 1944.

[114] Voytkovskiy, K.F., Eksperimental'nyye issledovaniya plasticheskikh svoystv l'da (Experimental Studies of Plastic Qualities of Ice), In book: "Seasonal Freezing of Soils and Utilization of Ice for Construction Purposes," Izd. AN SSSR (USSR Academy of Sciences Publishing House, Moscow, 1957.

[115] Shul'man, A.R., "Flowability of Polycrystalline Ice."
Tr. GGI, No 7 (61), 1948.

[116] Hess, H., "The Glacier" (Die Gletscher) [in German],
Braunschweig, 1904.

[117] Koch, K.R., "Concerning the Elasticity of Ice," [in German], Ann. der Physik, Vol 41, 1913; Vol 45, 1914,Leipzig.

[118] Brockamp, B., and Mothes, H., "Seismic Investigations on the Pasterze Glacier," [in German], Zeitschrift fuer Geophysik (Geophysics Journal), Vol 16, No 8, 1930.

[119] Tsytovich, N.A., and Sumgin, M.I., Osnovaniya medkaniki merzlykh gruntov (Fundamentals of Mechanics of Frozen Soils), Moscow, 1937.

[120] Glen, I.W., "The Creep of Polycrystalline Ice." Proc. Roy. Soc., Ser. A, Vol 228, N 1175, 1955.

[121] Proskuryakov, B.V., "Static Pressure of Ice on Structures, Tr. GGI, No 4, 1947.

[122] Berdennikov, V.P., "Study of Elastic Modules of Ice," Tr. GGI, No 7 (61), 1948.

[123] Monfore, G.J., "Ice Pressure Against Dams," Proc. of the American Society of Civil Eng., XII, Vol 78, Separate N 162, p. 13, 1952.

[124] Haefeili, R.,"Observation of the Quasi-Viscous Behavior of Ice in a Tunnel in the A'mutt Glacier,," Glaciology, Vol 2, No 12, 1952.

[125] Steinemann, S., "Results of Preliminary Experiments on the Plasticity of Ice Crystals," I. Glaciology, Vol 2, No 16, 1954.

[126] Fridman, Ya. B., Mekhanicheskiye sovoystva metallov (Mechanical Properties of Metals), Moscow, 1946.

[127] Shishov, N.D., "On the Strength of Ice," Meteorologiya i Gidrologiya (Meteorology and Hydrology, No 2, 1947.

[128] Butyagin, I.P. , "On the Strength of an Ice Pack During Bending," Tr. TEI ZSFAN

[129] Gumenskaya, O.M., Vliyaniye vlazhnostı i temperatury na soprotıvleniye merzlykh gruntov szhatiyu (Effect of Humidıty and Temperature on Resistance of Frozen Soils to Compression), 1936.

[130] Butyagin, I.P., "On Varıations in Thickness, Structure and Strength of Ice Sheet on Rivers of Western Sibéria Durıng the Spring (in Example of Ob' R.," Tr. III Vsesoyuz. digrol. s"yezda (Trans. of 3rd All-Unıon Hydrologic Congress), Vol 3, Leningrad, 1959.

[131] Shreyner, L.A., Tverdost' khrupkikh tel (Hardness of Brittle Bodies), Moscow-Leningrad, 1950.

[132] Zelenin, A.N., Fızicheskiye osnovy teorii rezaniya gruntov (Physical Bases to the Theory of Cutting Soils), Moscow-Leningrad, 1950.

[133] Surin, A.A., Vodosnabzheniye (Water Supply), Part I, Leningrad, 1926.

[134] GOST 3440-46. Loads, Ice.

[135] Dmitreva, N.A., Raspredeleniye napryazheniy v ploskosti sklayvaniya drevesiny (Stress Distribution in Shearing Plane of Wood), Kiev, 1948.

[136] Plaksin, M.V., Issledovaniye raskalyvaniya korotkıkh drevesnykh otrezkov (Study of Splitting of Short Pieces of Wood), L'vov, 1948.

[137] Buznikov, N.F., "On Level of Ice Pressure on a Bridge Pier," Gidrotekhnicheskoye stroitel'stvo, No 2, 1933.

[138] Afanas'yev, V.I., "Icebreakers," Articles published in newspapers "Kotlin" (No 49, 62,73 and 105 for 1897) and Kronshtadtskiy Vestnik (Kronstadt Herald),(No 123,124, 1895).

[139] Skobnikov, D.B., "Passage of Ice and Rubbish Through Structures of Volga HES imeni V.I. Lenin," Gidrotekhnicheskoye stroitel'stvo, No 1, 1961.

[140] Shimanskiy, Yu. A., "Provisional Methods of Measuring the Icebreaking Properties of Ships," Trudy Articheskogo in-ta (Trans. of Arctıc Institute) Vol 130, 1938.

[141] Tarshin, M.K., "Scientific Results of Research on Strength of Icebreakers," <u>Preolemy Arktiki</u> (Problems of the Arctic) No 2, 1938.

[142] Bernshteyn, S.A., <u>Osnovy dinamiki sooruzheniy</u> (Fundamentals of Structural Dynamics), Moscow-Leningrad, 1941.

[143] Runeberg, R.I., <u>O parokhodakh dlya zimnego plavaniya i ledokolakh</u> (On Steamers for Winter Cruising, and Icebreakers), St. Petersburg, 1890.

[144] Arnol'd-Alyav'yev, V.I., "Experiments on External Friction of Ice," <u>Zhurnal tekhnicheskoy fiziki</u> (Journal of Technical Physics), No 8, 1937.

[145] Korzhavin, K.N., "Action of Ice on Bridge Piers," Doctoral dissertation, V.I. Lenin Library.

[146] GOSSTROY USSR. Technical specifications for determining ice loads on river structures (SN 76-59), Moscow, 1960.

[147] Korzhavin, K.N., <u>K voprosu o velichine fakticheskogo davleniya l'da na opory mostov</u> (Contribution to Question of Actual Ice Pressure on Bridge Supports), Novosibirsk, 1958.

[148] Teleshev, V.I., Pinigin, M.I., and Tolokno, N.V., "Passage of Spring Ice Passage Through Structures of Mamakanskaya HES), <u>Gidrotekhnicheskoye stroitel'stvo</u>, No 7, 1961.

[149] Kazub, G. Ya., "Temperature Conditions of Ice Cover on Certain Rivers in Western Siberia," <u>Tr. TEI ZSFAN SSSR</u>, No 5, Novosibirsk, 1955.

[150] Korzhavin, K.H., "New Method for Determining Actual Pressure of Ice in Nature," <u>Tr. TEI ZSFAN SSSR</u>, No 7, 1958.

[151] Bibikov, D.N. and Petrunichev, N.N., <u>Ledovyye zatrudneniya no gidrostantsiyakh</u> (Ice Difficulties at Hydraulic Stations), Moscow-Leningrad, 1950.

[152] Latyshenkov, A.M., "Investigation of Ice-Shielding Panels, <u>Gidrotekhnicheskoye</u> stroitel'stvo, No 4, 1946.

[153] Bovin, V.T., "On the Question of a Wind-Caused Wave." Materials on Prof. I.G. Aleksandrov's project, No 3, Moscow, 1927.

[154] Komarovskiy, A.N., Zimnyaya rabota zatvorov gidrosooruz-
heniy (Winter Operation of Gates to Hydraulic Structures),
Moscow-Leningrad, 1933.

[155] Yestifeyev, A.M., "Control of Output at HES Cascade as a
Method of Reducing Construction Cost," Tr. TEI AN SSSR,
No 11, Novosibirsk, 1961.

[156] Gamayunov, A.I., "Pressure on Sloped Walls," Gidrotekhnicheskoye stroitel'stvo No 6, 1959.

[157] Korzhavin, K.N., Dinamicheskoye davleniye l'da sooruzheniya v usloviyakh ledokhoda rek Sibiri (Dynamic Pressure
of Ice on Structures Under Conditions of Ice Passage in
Siberian Rivers), Izd. VNIIG, 1957.

[158] Gamayunov, A.I., "Vertical Pressure of Ice During Variation
in Level of Ice Jamming," Gidrotekhnicheskoye stroitel'stvo,
No 9, 1960.

[159] Shankin, P.A., "Ice Effects on Concrete Coagings of Slopes,"
Gidrotekhnicheskoye Stroitel'stvo, No 9, 1960.

[160] Shankin, P.A., "On the Question of Strength Calculation of
Concrete Facing of Slope to Allow for Action on Ice Pripay
(Fast Shore Ice)," Gidrotekhnicheskoye stroitel'stvo,
No 3, 1961.

[161] Ioganson, Ye. I., Zimniy rezhim r. Volkhov i oz. Il'men'.
Mater. po issledovaniyu r. Volkhov i yego bassenya (Winter
Regime of Volkhov R. and Lake Il'men'. Materials on Study
of Volkhov R. and its basin). Leningrad, 1927.

[162] Kirkham, I.E., "Five Missouri River Bridges," Eng. News
Record, No 18, 1927.

[163] Zylev, B.V., Davleniye l'da na opory mostov (Ice Pressure
on Bridge Piers), Dissertation. V.I. Lenin Library, Moscow, 1948.

[164] Brown and G.C. Clarce, "Ice Thrust in Connection with
Hydro-Electric Plant Design," The Eng. I., January 1932.

[165] Ice Thrust Against Dams. Eng. News Record, Vol 99, No 19,
1926.

[166] VODGEO, "Gravity Dams," Provisional Technical Specifications and Standards for Planning and Erection, Moscow,1934.

[167]   Proyekty tekhnicheskikh usloviy i norm gidrotekhnicheskogo
        proyektirovaniya (Drafts of Technical Specifications and
        Standards for Hydraulic Engineering Planning, Gosstroyizdat,
        Moscow-Leningrad, 1939.

[168]   Orlov, F.F., "Bending Strength of Ice on Oka R.," Tr.
        Gor'kovskogo in-ta inzh. vodn. tr-ta. (Transactions of
        Gor'kiy Institute of Engineering and Water Transport),
        Vol 7, 1940.

[169]   Pisarev, P.A., "Flexural Strength of Ice," Krasnoyarsk,
        1944.

[170]   Shishov, N.D., Zheleznodorozhnyye perepravy po l'dy cherez
        r. Severnuyu Dvinu u Arkhangel'ska (Railroad Crossings Over
        the Ice Across the Northern Dvina at Arkhangel'sk),
        Arkticheskiy Institut (Arctic Institute), 1945.

[171]   Komarovskiy, A.N., Struktura i fizicheskiye svoystva
        ledyanogo pokrova presnykh vod (Structure and Physical
        Properties of Ice Sheet over Fresh Water), Gosenergoizdat,
        Moscow-Leningrad, 1932.

[172]   Proskuryakov, B.V., "On an Analysis of the Drift of Ice
        Fields," Problemy Arktiki, No 5, 1941.

[173]   Petrunichev, N.N., and Mamayev, I.M., "Example of Cal-
        culating Velocity and Direction of a Drifting Ice Field,"
        Problemy Arktiki, No 5, 1941.

[174]   Ofitserov, A.S., "Pressure of Broken Ice Field on Struc-
        tures," Gidrotekhnicheskoye stroitel'stvo, No 9, 1948.

[175]   Polukarov, G.V., "On Calculating Ice Pressure on Hydraulic
        Structures," Tr. Gos. Okeanogr. in-ta. (Transactions of
        State Oceanographic Institute, No 23, 1953.

[176]   Latyshenkov, A.M., "Investigation of Ice-Protecting Panels,"
        Gidrotekhnicheskoye stroitel'stvo, No 4, 1946.

[177]   Zubov, N.N., "Concepts on Movement of Ice Under the Effect
        of Wind," Issl. morey SSSR (Investigation of USSR Seas),
        No 21, 1935.

[178]   Yegorov, A.I., "Novosibirsk Hydroelectric Station on Ob'
        R.," Gidrotekhnicheskoye stroitel'stvo, No 1w, 1957.

[179] Feriger, B.P., "Krasnoyarsk HES on Yenisey R.," Gidro-
       tekhnicheskoye stroitel'stvo, No 5, 1961.

[180] Vasıl'yev, A.F., "Passage of Ice Through Hydraulic Units,
       Gidrotekhnicheskoye stroitel'stvo, No 1, 1958.

[181] Davidenkov, N.N., "On the Question of Undertaking Experi-
       ments for Measuring Ice Pressure," Sb. po ledotekhnike
       (Collection on Ice Engineering), No 1, Moscow-Leningrad,
       1933.

[182] Panov, B.P., "Ice Dynamometer," Problemy Arktikı, No
       10-11, 1939.

[183] Plakida, M.E., "Problems on Studying the Ice Regime in
       Reservoirs," Tr. III Vsesoyuz. godrol. s"yezda, Vol 3,
       Leningrad, 1959.

[184] Konavalov, I.M., and Makkaveyev, V.M., Gidravlika
       (Hydraulics), Rechizdat, 1940.

[185] Konovalov, I.M., "On the Velocıty of a Drifting Body in
       a Flow," Tr. LIIVT, No 5, 1934.

[186] Belokon', P.N., Inzhenernaya gidravlika potoka pod
       ledyanym pokrovom (Engineering Hydraulics of Flow Beneath
       an Ice Pack), Moscow-Leningrad, 1940.

[187] TSNII lesosplava (Central Scientific-Research Institute
       on Timber Floating), No 97,98, Moscow, 1938.

[188] Zvonkov, V.V., Tyaga rechnykh sudov (Thrust of Riverine
       Vessels), Moscow, 1940.

[189] Shlikhting, G., Teoriya pogranichnogo sloya (Boundary
       Layer Theory), Moscow, 1956.

[190] Shuleykin, V.V. Fizika morya (Marine Physics) Moscow,1941

[191] Bydin, F.I., "Development of Certain Questions in the
       Field of Ice Regime in Basins," Tr. III Gidrol. s"yezda,
       Leningrad, 1959.

[192] Runeberg, R., and Makarov, S., "On Construction of Ice-
       breakers," Morskoy sbornik (Marine Collection), No 10,
       1898.

[193] VNIIG. Suggestions for review of GOST for ice actions.
       Leningrad, 1959.

[194] Morgunov, V.K., "New Technique for Determining the Velocity
       of Floes' Movement and Their Dimensions by Photogrammetry
       Procedures," Tr. TEI SO AN SSSR, No 11, Novosibirsk, 1961.

[195] Morgunov, V.K., "Simplified Method for Determining Sizes
       of Floes Based on Photographs," Izv. SO AN SSSR, No 7,1960.

[196] Veselovskiy, N.I., Fotogrammetriya (Photogrammetry),
       Geodezizdat, Moscow, 1945.

[197] "International Association on Hydraulic Research. Ice
       Problems in Hydraulic Engineering Structures," Seminar
       No 1, 8th MAGI Congress, Montreal, 1959.

[198] Barnes, Kh., Ledotekhnika (Ice Engineering), Leningrad,1934

[199] Pospelov, B.V., "Passage of Flood Stages, Ice Passage of
       Timber Floating in Chennel of River Confined by Cofferdams.
       Gidrotekhnicheskoye stroitel'stvo, No 1, 1956.

[200] Pospelov, B.V., "Construction and Operation of a Large
       Cofferdam," Gidrotekhnicheskoye stroitel'stvo, No 3,1955.

[201] Ivanov, K.E., and Lavrov, V.V., "Concerning One Feature in
       the Mechanism of Plastic Deformation of Ice." Zhurnal
       Tekhniceskoy Fisiki, Vol 20, No 2, 1950.

[202] Voytkovskiy, K.F., Raschet sooruzh eniy iz l'da i snega
       (Design of Structures Made of Ice and Snow), Izd-vo AN
       SSSR, Moscow, 1954.

[203] Al'tberg, V.Ya., "Forces of Ice Freezing Together with
       Certain Substances," Tr. GGI, No 4(58), Leningrad, 1958.

[204] Rymsha, V.A., Ledovyye issledovaniya na rekakh i vodokh-
       ranilishchakh (Ice Studies on Rivers and Reservoirs),
       Gidrometeoizdat, Leningrad, 1959.

[205] Korzhavin, K.N., "On an Efficient Type of Bridge Pier Under
       Conditions of Ice-Out in Siberian Rivers," Tr. NIIZHT,
       No 7, 1949.

[206] Bydin, F.I., "On Principles of Combatting Ice Passage,"
       Tr. TEI SO AN SSSR, No 11, Novosibirsk, 1961.

[207]   Agalakov, S.S., "Passage of Output of Water and Ice During
        Construction of a Hydroelectric Center with High Concrete
        Dam on River with Plentiful Water Supply," In the collec-
        tion: "Planning and Construction of High Dams." Moscow-
        Leningrad, 1960.

[208]   Yestifeyev, A.M., "Regulation of Ice Passage During Period
        of Erection and Operation of Dams Under Conditions of
        Bratskaya and Krasnoyarskaya HES." In the collection:
        "Planning and Construction of High Dams", Moscow-Lenin-
        grad, 1960.

[209]   Shadrin, G.S., and Panfilov, D.F., "Dynamic Pressure of
        Ice on Hydraulic Structures," Izv. VNIIG, No 69, Moscow-
        Leningrad, 1962.

LIST OF CERTAIN ABBREVIATIONS ADOPTED IN THE BIBLIOGRAPHIC REMARKS

TPEI - Transport Power-Engineering Institute of the Siberian Branch of the USSR Academy of Sciences.

AUSRIHEiV - All-Union Scientific-Research Institute of Hydraulic Engineering imeni B. Ye. Vedeneyev

SHI - State Hydrologic Institute of GUGMS

NIRTE - Novosibirsk Institute of Railway Transport Engineers

MIRTE - Moscow Institute of Railway Transport Engineers

LIWT - Leningrad Institute of Water Transport

AUSRIRT - All-Union Scientific-Reserach Institute of Railway Transport

CFI - Central Forecasting Institute of GUGMS